Manual of Construction
Project Management
for Owners and Clients

Manual of Construction Project Management
for Owners and Clients

Jüri Sutt

Professor of Construction Economics and Management
Tallinn University of Technology

A John Wiley & Sons, Ltd., Publication

This edition first published 2011 © 2011 by John Wiley & Sons, Ltd

Blackwell Publishing was acquired by John Wiley & Sons in February 2007. Blackwell's publishing program has been merged with Wiley's global Scientific, Technical and Medical business to form Wiley-Blackwell.

Registered office:
John Wiley & Sons, Ltd, The Atrium, Southern Gate, Chichester, West Sussex, PO19 8SQ, UK

Editorial offices:
9600 Garsington Road, Oxford, OX4 2DQ, UK
The Atrium, Southern Gate, Chichester, West Sussex, PO19 8SQ, UK
2121 State Avenue, Ames, Iowa 50014-8300, USA

For details of our global editorial offices, for customer services and for information about how to apply
for permission to reuse the copyright material in this book please see our website at www.wiley.com/wiley-blackwell.

Library of Congress Cataloging-in-Publication Data

Sutt, Jüri.
  Manual of construction project management for owners and clients / Jüri Sutt.
    p. cm.
  Includes bibliographical references and index.
  ISBN 978-0-4706-5824-6 (pbk. : alk. paper) 1. Building–Superintendence–Handbooks, manuals, etc. 2. Project management–Handbooks, manuals, etc. 3. Contractors' operations–Handbooks, manuals, etc. I. Title.
  TH438.S8879 2011
  690.068–dc22

                           2010051096

A catalogue record for this book is available from the British Library.

This book is published in the following electronic formats: ePDF 9781119971689; ePub 9781119971696 and MobiPocket 9781119971702

Set in 11 on 14 pt Palatino by Toppan Best-set Premedia Limited
Printed and bound in Malaysia by Vivar Printing Sdn Bhd

1   2011

Contents

Preface

Each building project requires project management from three different perspectives: that of the client, the designer and the builder. Although they will share a common knowledge of project management, they will also have specific knowledge of their own fields of management. This book describes the activities of the owner in the role of construction client. By performing the complete list of activities himself, the owner maximises the probability of achieving his quality, cost and time objectives. The owner must know which risks he will accept when leaving some of the activities to his partners. Describing the owner's activities from the initial idea to build right through to final readiness, the book is a manual for the owner. At the same time the client's partners must know what the client expects from them; so this book is addressed to designers, contractors, supervisors and professional construction managers as well. Within this text, the owner's activities preceding design, construction and design procurement are particularly highlighted. As nearly half the construction managers and economists in the construction market are engaged on behalf of the owner, the book can be used as study material for construction-faculty students.

The author would like to thank Erki Laimets, director of Conviso Ltd., whose suggestions for the manuscript's structure, and numerous remarks thereon, were gratefully received. The author also gratefully acknowledges the remarks and review of Lembit Linnupõld, director of Estkonsult Ltd., and the help of Ahti Väin, director of Ahti Väin Konsult Ltd., Vladimir Issakov, director of Tallinn Linnaehitus Ltd., and Jaanus Tehver, attorney-at-law with Tehver and Partners. The efforts of Diana Järve and Daniel Edward Allen in translation and editing are also highly appreciated.

About the Author

Jüri Sutt has nearly 50 years of experience in construction management as a practising manager, researcher, consultant and lecturer, which has included designing the construction technology for large mines in Siberia and a gas trunk pipeline in Libya, and also managing a construction company. In 1965 he pioneered the use of IT in construction-management research in Estonia. Between 1965 and 1980, Jüri Sutt was a member of several USSR scientific councils in the field of construction management and, from 1965 to 1978, was the head of the Construction Management Department of Estonia's State Building Research Institute which developed scheduling and cost-estimating IT systems that were widely used in the Soviet Union.

He has been an adviser to four ministers responsible for building during Estonia's transition to a free-market economy and led working groups formulating construction-market regulations in the 1990s. In addition, he has provided consultancy services for clients' projects and contract management and has gained expertise in contract disputes in the last 15 years.

In 1960, Jüri Sutt qualified as a construction engineer. He was awarded the Candidate of Science degree in 1968 (equivalent to a PhD), and, in 1989, the Doctor of Science (habil.) in mathematical methods and IT in economics. The principal outcome of his research has been the methodology of IT simulating production – economic activities of construction companies enabling experimentation with different economic mechanisms and management strategies in construction enterprises.

Since 1989, he has been Professor of Construction Economics and Management at the Tallinn University of Technology.

Introduction

All of us as individuals, families or organisations are either owners of buildings or parts of them, or are at least the occupants of them. In developed countries, the building sector accounts for about 7–10% of GNP and the process of construction is expensive and time-consuming. Each prospective owner has his individual needs and demands, and expects the building to satisfy them. The state, through its laws and decrees, sets down rules that must protect social interests, public safety and the economic use of resources. Local governments must ensure, through their own regional plans, laws and rules, rational and cooperative, conflict-free regional development in the long term. The public-orientated character of local planning protects the interests of neighbours and local people when any new construction work begins. Within the construction industry, various enterprises and professional associations have laid down standards, directives and general conditions relating to contracts, classifications of works and resources, construction costs and design documents. The acceptance of such standards by the interested parties can be seen as part of the written rules of good practice within construction.

Every building is complex both in the way it will be used and in its structural composition. Every building's owner must negotiate many stages in the building's life cycle before completion: determining (formalising) his needs, profitability and feasibility studies (investigations), financing schemes, purchasing the land, detailed planning of land use, and, only then, the design and construction of the building. Today the owner as a rule does not directly manage the construction himself. These activities are divided between many agents linked by contractual relationships: land-use consultants; cost engineers; investigators of ground, of water, of neighbouring buildings and of rights and burdens of ownership; designers; main-, sub- and prime-contractors; materials' suppliers; financiers; insurers of different risks; local or national authorities, etc. To coordinate these activities through all of the development stages, and through the stages of construction itself, a specific project-management function is required.

In planning contractual relationships and administering the project, the future owner operates in the role of client. Prospective building owners are usually professionals in other areas of business and act as construction clients very seldom. Moreover, they do not have the ability or resources to success-fully fulfil the role of building client themselves. For this reason, professional construction management for the owner has developed within the construction industry over the last few decades. Nonetheless, it should be said that to achieve the best results the prospective owner must also participate in the man-agement process.

The three most important criteria must remain under the control of the owner throughout the life cycle of the building: **cost, quality and time**.

During the first stage of any project – the **project prepara-tion stage** – the main problems associated with costs are forecasting and planning the cost and the revenue of the build-ing, the profitability of the project, balancing the costs of con-struction and the costs of using the building, and establishing the project's overall quality and the indices for control to achieve it.

During the second stage of building, **design** is the main activ-ity. Experience in construction demonstrates that, at this stage, owners often pay too much attention to the cost of design, at the expense of paying the necessary attention to the control of construction costs. For example, they might choose a designer only because he quotes the lowest price; or neglect to obtain bills of quantities with detailed descriptions of the specifications for the construction works, materials and control estimates; or they might not stipulate the responsibilities for controlling construction costs in the design consultant's contract. It should be remembered that the results of the design stage determine the project's future overall costs.

At the design stage, by neglecting to include descriptions of quality and calculations of quantities, and by omitting to describe costs with enough detail, the owner could save 0.1% of the project's construction cost, although this would leave many future choices to construction contractors; however, it should be remembered that poor estimations of material requirements and poor descriptions of the required quality can become the source of future disputes.

In fact, the failure to produce accurate specifications and quantities of materials and works together with an accurate cost forecast is the most common reason why an owner's expectations are not matched by the results, and why the actual cost of these results can substantially exceed the budgeted costs.

At the third stage – the **stage of construction** – there are many different methods of cost management. By choosing one of them, the owner must attempt to find a balance between maximising benefits and minimising risk. During the last few decades the main developments in the construction market have come from pressure to shorten the duration of the construction period. The substantial shortening that has been achieved has come not from the shortening of construction site works, but rather by combining the timing of the client's project-management activities, as it has proved effective to combine the methods of building procurement. Any shortening of construction time on the building site, by using two or three shifts or by using more powerful equipment, can cause a rise in costs for the contractor, which must be paid by the client. The duration of on-site construction must be approximately determined during the preparation stage, in the phase of profitability studies, by consulting with contractors or professional construction managers. Construction duration makes up the substantial part of the project's duration and therefore has a great influence on profitability, financial plans and the overall cost of construction. The decision on construction duration is the basis from which to determine the contract's start and finish dates, which must be clearly identified in documents when invitations to tender are issued.

It must be said that maximising the client's outcome on all three criteria – cost, quality and time – is not possible. Saving on the cost can affect quality, as shortening the duration of construction might increase overall costs as well as increasing the project's risk and downgrading the quality.

This book is a construction client's manual, written in the form of a list of activities, with the aim of guiding the owner in the role of client. The book will help clients to make choices during the project-development process, guaranteeing control over cost, quality and duration at each stage, taking into account the individuality of projects and owners. Activities within each stage are divided into phases, each of which requires separate decision-making. Each phase begins with a list of preceding decisions and the goal of the present phase, continues with a list of activities to be performed and description of the roles of their executors. Each phase ends with a list of expected results and a list of activities that these results enable to be actioned in the next phase.

Activities in this book are presented according to the main stages of project development: the preparation stage, the procurement stage, the design stage, preparation for construction, construction itself, the hand-over stage and the stage of implementation. The sequence of these stages can be altered because one way in which a prospective building owner can manage his risk is by choosing and combining the timing of these stages. The tasks involved in project preparation, described in the first chapter, are therefore often left by the owner for the designers to perform, or sometimes even for the contractors. The decisions relating to the choice of procurement schemes, described in the second chapter, can be made either at the preparation stage of project development, as part of the prioritisation of aims, or at the time of choosing the designer, or at the stage of choosing construction contractors.

It is rational to include a description of procurement as a separate stage in the project's life cycle, as, in procurement schemes, preparation for the development, design and construction stages is inseparably connected and often operates simultaneously, and the methods of implementation are often the same. The preparation of procurement is frequently a separate service, supplied by the professional construction manager or the project's principal adviser. One of the reasons why this book discusses procurement schemes and the motives for the choice between them in a separate chapter is to avoid duplication in other chapters.

At the same time there is some duplication in the list of activities because of the different possibilities for project development: for instance, feasibility studies and the purchase of land, along with the preparation for purchase activities, depending on whether the client has to buy a plot for the planned building or if he owns it already.

Appendix 1 summarises in list form the main decisions that the client must make during the development of a project, while Appendix 2 gives a list of the folders that it will be necessary to create and to store. Of course, in different countries differences in culture and in 'construction best practice' may influence the client's decisions, although the generalities of possible schemes are the same in all market-economy countries.

As it is always easier to avoid the unnecessary than to discover the unknown, the list of client activities in this manual is presented in its fullest form – meaning that not all of them will be necessary on every project. The activities are not divided between main and secondary activities, as the importance of each activity will be governed by the demands of each project, or each stage within the project. It can be said that by following the listed activities, there will be a greater probability of achieving maximum economy in building costs, shortening the duration of the project and ensuring the success of the client's targeted quality measures. It is necessary to be aware that the owner's costs increase in sequence from stage to stage, while with each successive stage the influence on the outcome of the project (cost, quality and duration) decreases.

The prospective owner's activities in procuring a new building can follow the different schemes described in this manual, with different schemes allocating different contractual partners different proportions of risk. The manual cannot give recommendations for specific projects, but it does describe possible motives for choosing one or another alternative. The prospective owner is the main decision-maker when allocating the risks involved in a project's development – the client is 'king'.

In order to save time or money (in relatively small amounts at the beginning of a project), the owner can choose not to consider alternatives in scheme design; or omit any of the following: putting together bills of quantities or detailed specifications and control estimates; using a principal adviser for planning and operating procurement; ordering special quality plans, etc. However, the owner must understand that by omitting these activities he increases his risks. This is the reason why we often read in the press that the actual cost of a building is higher, and the actual quality lower, than expected. And no consumer protection law or court case can help if the necessary preparatory activities were not undertaken and their results included in the contract documents.

The manual describes activities at the level of detail required to choose the management task or method to make the decision, while omitting the decision-making procedures themselves and the specific forms of the documents.

The manual is addressed to prospective owners who either operate as clients themselves, or who use the services of professional construction-management companies. The aim is to help owners understand what their construction partners expect from them.

The manual does not include special activities and rules for the management of public procurement, which are determined in the relevant legal documents. Neither does the manual describe specific rules for contracts in the sectors of road and railway construction, or for construction of other special technical facilities.

To ensure unambiguous understanding of the project-development process, the manual begins with an explanation of terms. Instead of appearing in alphabetical order, terms are presented from general towards specific and detailed. This should assist with the understanding of the client's decision-making process in managing his project in the building market. Numerical appraisals in the text of the costs or timespan of described activities are averages derived from experience. In each case the owner must take into account the specific details of the project and the environment including local legislation. However, the presented appraisals can help the inexperienced client in his first steps.

The terms in italic in the manual should be understood according to their meaning given in the Glossary. The use in the text of 'him', 'his', 'himself', etc. is intended to apply equally to persons of either gender and is employed only for convenience in writing.

Glossary

The terms necessary to describe economic, management and procurement objectives and processes can have different meanings in different countries depending on cultural differences. Because the use of terms in official documents could lead to an economic and juridical outcome, it is advisable to consult with the contract's principal adviser in cases of doubt. The terms explained below appear in italic in the main text of this book.

Building, design and construction – these words as terms must be used in legal documents in accordance with building law, which varies from country to country. In this manual a building as a product can consist of structures, or communications and other technical facilities associated with them. A building as a physical body must be permanently attached to land. If the term 'project' is used to mean investment in construction then it can include the formation of several buildings.

In this case, buildings within the scope of the project are defined as objects of construction. **Building objects** (structures, facilities and their groups) can form independently implemented complexes, and can use different financing schemes, while the level of development preparation can also be different.

Price and cost – used as parallel terms, depending from which side of the partners the notion is described.

Good construction practice – a framework consisting of individual practices and procedures, on which the collaboration of contract partners, the authorities and other interested parties is based, and which is used in order to ensure that the client's objectives for building cost, quality and time are met. Following good practice ensures that all parties take each other's interests into account in a fair and unbiased fashion, as well as solving disputes within a reasonable time period. Good construction practice is constantly developing, as are the requirements of construction quality. Good construction practice is formed from rules and conditions determined within the general conditions of contracts, instructions and standards that are publicly approved by associations of partners' enterprises and professions in order to ensure mono-semantic understanding, and a clear and transparent sharing of the risks between partners. It is in the nature of most good-practice rules that they are not obligatory, so it is advisable to stipulate their use in the contract documentation. If contract conditions are not described clearly enough in these documents, then cases of dispute will be resolved in the courts based on the use of good construction practice. (At least, we hope!)

General conditions of contracts (GCC) – are composed with the aim of reducing labour intensity (a general contract between owner and builder for ordinary building includes some tens of pages). Harmonised by associations of entrepreneurs and trades representing all sides of the contract, the GCC guarantee the potentially equal division of risk between the parties involved. For owner (client), as for other parties with lesser experience in construction, using the GCC is especially important. General conditions of contracts are usually specified in all countries and are slightly different depending on local laws, traditions and culture. General conditions are usually used for contracts between the owner and the general contractor; the owner and the designer; the owner and the supervisor; the owner and the professional construction manager; and between the general contractor and the subcontractors. For small building projects and 'single construction' works, special GCC would be used. As GCC are not mandatory, it is necessary to add them as an appendix to the contract thereby making them binding for both parties.

Reconstruction – rebuilding that changes the main structure or the function of the building.

Renovation – rebuilding that does not change the main structure of the building, but which increases the value of use of the building without changing its function.

Repairs – construction works to repair or build worn-out or broken building or facility elements, including structural repairs to restore parts of the main structure and the ongoing maintenance repairs required for the normal functioning of the building and to extend its useful life.

Owner – the principal of the building-investment project and the owner of the building once completed. During the project, the owner's activities depend on whether he is building to occupy the finished building, to rent it out or to sell it. In his dealings with contractors (builders, designers, consultants, etc.), the owner acts in the role of **client**. He can fulfil the role of client himself or delegate it, by means of a contractual relationship, to professional construction managers. The activities allowable for a public owner are determined by law.

Plot of land – a plot of land in this context is determined according to the *detailed area plan* to be used for building. During the construction period it, or some part of it, becomes the building site.

Detailed area plan – project documents (specification and drawings) resulting from the design that determine the purpose and conditions of use of the land, restrictions for buildings on it, as well as dividing the site into real estates, determining conditions for traffic, landscaping, technical facilities, servitudes, etc. As all aspects of area planning and the debate on the results are public, and end with the approval of local government, it can be looked on as public agreement for land use, protecting the interests of neighbours and of society. Conditions determined in the detailed area plan are legally binding.

Servitude – constrained right to use either your own land, or that belonging to another (necessary, for example, for building a water-supply pipeline through a neighbour's plot of land). The conditions of the constraints, detailing the rights of both parties, should be stipulated in servitude contracts.

Building site – (construction site) is an area of ground (or, indeed, possibly under or above it) on the *plot of land*, or at another location in the case of storage or prefabrication, that the owner has appointed for the construction work. During construction the owner gives possession of the building site to the main contractor, at the same time making him responsible for it. It should be noted that the plot of land and the building site might not be identical areas of land for the reasons mentioned above.

Project – is a term that can be used in several languages with different meanings. The first, wider meaning is that of a set of activities geared to achieving an agreed outcome with criteria to measure the results. For each project, the beginning and the end are fixed. Executing the project presumes planning (the allocation of resources between the activities of the project). An example of a project would be the owner's planned building-development project (investment project) beginning with profitability and feasibility studies, continuing with the construction period and finishing with the use and maintenance of the building and ultimately with its possible demolition. The criterion for evaluating the project could be, for example, the payback period.

The term **project** is also used in the context of building to mean the result (product) of the designer's output, which is to say the package (complement) of technical and cost-related documents (descriptions, drawings, calculations, specifications, etc). This meaning of 'project' is the prescriptive model of building. The design process is divided into stages, making it possible for owners to control the process and minimise risks, and for the relevant authorities to evaluate the building before granting a licence to construct. Project documents must describe the future building in such a way that the builder can meet the expectations of the owner. Usually every country has its own standards describing the required content and level of detail of different project stages.

Project management (using the first definition of *project*, above) includes the following functions: determination of scope; appointment of a project manager (PM) and integration of the PM into the process management of the company; cost management; quality management; project-time management; resource management; management of communication; risk management; conflict management.

Scheme design (brief) – preceding the detailed and complex design phase of project development including results of investigation of needs, profitability and feasibility studies with scheme drawings. Not complex in this context means that scheme design does not include treatment of all parts of the project in the meaning of design (although rough).

The aim is to balance the needs arising from production technology or from the requirements for the amount of space with the restrictions of the environment (detailed planning and architecture) and the limits of finance (the financial return of the project). Before starting detailed design (except in small-scale buildings such as private residences), it is necessary to answer many technical, economic and organisational questions in order to aid decision-making and for the approval of subsequent steps in the building-investment project. For this reason calculations from the preparation stage (scheme drawings, cost calculations, quality needs, time-requirement appraisals and preliminary procurement schemes) are assembled in a single package, which from the owner's viewpoint effectively forms the scheme design. The content and detail of the scheme-design package must be adequate to allow decisions on the expediency of investment and land purchasing, to begin detailed planning procedures and to enable any relevant negotiations with the local authorities. Whether the owner himself or whether his consultants prepare the documents is less important. What is important is that the person who will be responsible for putting the package together is appointed in a timely fashion. As the project develops, it will become, step by step, more complicated and the owner will not have the ability and special equipment necessary. Nowadays scheme drawings can be prepared using CAD methods. It is therefore prudent for the owner to involve a cosultancy company for scheme design. This will, at the same time, decrease the cost of more detailed and complex preliminary design for the owner.

Scheme – a sketch of the principal solution. For a simple building (e.g. a family house), scheme design can be limited to architectural sketches, a sketch plan of the site and notes about the required quality standards. At this stage the cost estimates are based on market prices.

Preliminary design – the first phase of more detailed and complex design, in which the principal technical solutions for the building are provided in drawings and in descriptions for all parts and facilities therein. The content is enough to enable an approximate cost appraisal, as well as obtaining the approval of the owner and agreement from the necessary institutions, including permission to build from the local authorities. However, it is not detailed enough that construction can begin. In this context preliminary design and preliminary project are synonymous.

Basic design – adds specific detail to the data of the preliminary design. This includes drawings with all necessary measurements, bills of quantities and specifications describing the quality of materials and works. Solutions are detailed enough that construction can be based on them. It does not determine construction technology or materials' manufacturers. It is usually the client who makes the commitment to the basic design.

Working drawings – detail and specify solutions presented in the *basic design*. Commitment to the working drawings comes from the construction contractor, or – more rarely – the client. Working operations on the construction site are executed in accordance with the basic design or working drawings. Approval of working drawings by the designer may be required. There may be cases when working drawings are not necessary, for example, when the contractor is experienced in the particular technologies involved. The owner's supervisor has the authority to request working drawings to be prepared for specific technological elements of the building.

Maintenance and operating instructions – guidelines for a building and its facilities that include plans and instructions for equipment, as well as schedules for regular maintenance, etc.

Demolition design – prepared at the level of detail of working drawings' design and includes the sequence of activities, the required temporary shoring of elements, the approximate amounts of demolished materials and proposals for waste management.

Performance design (project) – produced during construction to rectify any differences between executed elements and facilities and those as designed.

Surveying drawings – taken from measurements of completed buildings, needed mainly for legal tasks connected with registers and property management.

Builder – in the context of building procurement the builder is the **contractor** and the term can refer to the general (main) contractor or a specialised subcontractor. The main contractor is a corporate or physical body selected by the client and empowered through a contractual relationship to execute the works himself, or to execute the works by dividing them between other (sub)contractors. The general contractor is responsible to the client for all work carried out by the companies subcontracted to him. In addition to the general contractor, the client can choose and empower other contractors to perform particular sections of work, in which case these contractors are responsible directly to the client. According to *good construction practice*, every country has detailed general conditions for contracts describing the division of roles, responsibilities and risks between the client and his contract partners. Some countries require the official registration or licensing of builders as contractors.

Designer (consultant) – for a client this generally means the main contractor for design and consultancy work: a company that fulfils this particular role alone or with the help of subcontractors. As in the case for the builder, the general contract conditions between the client and consultant (designer) are detailed, in addition to which, registration and licensing may be required. *Direct contractors* are rarely used in building design because the coordination of the separate parts of the project, specified in the drawings and other technical descriptions, is hard to achieve without a general design contractor. Basing his actions on *good construction practice*, the designer acts in accordance with the general conditions established between the owner and designer.

General contractor (main contractor) – using the fixed, lump sum contract, the general contractor takes full responsibility for price, quality and deadlines. If guarantee of price and quality is important for the owner, then it is necessary to ask for a tender price using tender documents based on the basic design, including **detailed bills of quantities** and **descriptions (specifications) of the work** (standards of quality). An accelerated option of general contract is also possible, when a request for the cost is made before the basic design stage is finished. In this case, it is possible to use price mechanisms which include cost reimbursement, or unit-price methods.

Direct (prime) contractor – a corporate or physical body that is selected by the client and that works in parallel with the general contractor. Dealings between direct contractors, the client and the general contractor, including subordinate activities on the building site, are established through contractual agreements.

Design and build contractor – takes responsibility for cost, quality and deadlines during both the design and construction stages of the project. Usually for industrial buildings, the technological part of the design is needed before the procurement of construction design can begin. The design of production technology in the preliminary-design stage and the design of the building from the scheme-design stage must be completed. For civil buildings (housing, offices, etc.), the procurement procedures can begin based on scheme design, if the owner takes the risk of being granted permission for the building. In comparison with the main contract, the owner accepts much more risk. The advantage of doing this is that it can shorten the duration of the project due to a single company performing the functions of design and construction, in addition to which, construction experience at the design stage is also beneficial.

Professional construction manager – the building's project manager who fulfils the role of construction client according to the mandatory contract, when the owner himself lacks the time and/or ability. Management services can include all stages of the project, beginning with scheme design and investigations and ending with the services at the guarantee stage of building, or they can include a lesser number of stages or phases. In contrast to the general contractor, the professional construction manager does not take responsibility for the cost of construction, and neither does his profit depend on the difference between the price and the cost of construction. The result of his work is not a product (the building) but management services. The construction manager prepares the procurement procedure as well as the contracts with all contractors, although the owner himself will sign the contracts. When using a professional construction manager, the price of the building will be clear once the last contract is signed. This method in its pure form uses neither main contracts nor subcontracts, but rather direct contracts between the owner and all prime contractors. To ensure the lowest price for the owner, direct contractors must be found through lowest-price competition. The expertise of the professional construction manager acts as a guarantee of good quality in construction.

Owner's construction supervisor – responsible for the control of construction work according to design and contract documents, legal norms and the regulations of local authorities. The owner's supervisor operates in accordance with the *mandatory contract* with the owner.

Building site diary (log) – contains daily notes on the main events and conditions of work on the site; the diary may affect the result of the project in the future. Completed in one folder, retained together with other contract documentation, it can be used as one of the main documents in cases of contract dispute.

Act of covered works – a document dealing with works that will be covered by other parts of the building. The act of covered works requires an inspection before the covering-up, agreed by the manager of the construction works and the owner's supervisor.

Principal adviser – an adviser from outside the owner's company responsible for recommending the choice of procurement schemes, tendering procedure, contract-price mechanisms, contract conditions, etc., including recommendations concerning the use of a professional construction manager. Obviously, the role of principal adviser should be taken by a highly experienced and qualified person. In comparison with the professional construction manager, he must be familiar with solving key problems; for assistance, he uses personnel from the owner's company. This does not exclude the professional construction-management company from acting in the role of principal adviser.

Building procurement – the sharing of the roles, rights, liabilities, responsibilities and risks in the building's life cycle between all contracted parties in order that aims in cost, quality and time targeted by the owner are achieved in the best fashion. This will be ensured by using procurement schemes, types of contracts, pricing mechanisms, tender procedures, etc., all of which are appropriate to the individuality of the building and the building's owner.

Competitive bidding (tender) – a procedure to select the most appropriate partner for a contract. In the construction market, the contractor is usually selected using the criterion of the lowest price, although time, quality and operating costs can all be taken into account.

Applicant – a company making a proposal to execute construction or design work in the first stage of two-stage tendering.

Bidder – a company making a price proposal.

Design contest – a method by which the client can purchase a design, sketch, etc. selected by a panel, mainly in the area of architecture, planning or structural design. The purpose of the competition is either to enter into a contract with the winner, or simply for the client to find conceptual solutions; the winner receives an award or is compensated for participation in some other way.

Partnership – a method of procurement in which contract partners are not found through tender but through previous working relationships based on positive experience. The advantage is that trust is already established, as are controlled management technologies and two-way confidence, bringing the possibility of price benefits. Procurement partnership schemes can be used between client and contractor, as well as in main contractor–subcontractor chains.

Procurement contract – in this context a contract, the object of which is construction or design work, or both. The procurement contract deals only with the result of the work to be carried out, not with the process of that work.

Framework (umbrella) contract – a contract between the owner and one or more builders to construct one or more buildings. The framework contract establishes common contract conditions for all future contracts.

Concession contract – a contract according to which the owner pays for building construction by giving the right to exploit the building in some way, or with this right and a concomitant financial payment.

Mandatory contract – in the context of building procurement, a contract in which the owner delegates his role of management, or his role of building supervisor, or both, to the contractor (a corporate or a physical body). The object of the contract is not a product but a service.

Bidding-invitation documents – documents necessary to invite interested parties to participate in competitive bidding or in negotiations. The documents include:

❑ advertisement and/or invitations to participate in *tendering*, containing short descriptions of the project;

❑ package of project (design) documents for the building (to achieve the lowest level of client risk, the bills of quantities and specifications should be included);

❑ programme of contract: the conditions of the contract that influence construction costs, which, because they influence cost, must be known by bidders for cost and time estimation.

Bills of quantities – quantities of works described at a level of detail that includes unit prices; prescribe the scope of the project.

Usually every country has a catalogue of average unit prices of works used for cost estimation before a building contractor is selected. The descriptions of unit prices are coded and as there can be more than one catalogue, the name of the catalogue must be given. The descriptions of works and prices vary between catalogues, but the level of detail is quite similar. The total number of unit prices in general construction works in all countries is 15–20,000. As construction enterprises are usually specialised (corresponding to the type of buildings or works), the works with their own workforce need fewer than a thousand unit prices for their description and pricing.

For cost management on behalf of the owner, it is preferable for the control estimate to be procured from a designer or cost consultant. This makes it possible to compare the bids of different bidders and to avoid disputes during the process of future project development. The alternative is to request the bills of quantities from the bidder. In the bills, works must be grouped by the classification standard of aggregated building elements or construction works. In the bidding process, bidders make their own cost estimates according to the bills of quantities.

Bill of activities – the list of the works required at the level of building elements or aggregated works (for example, external walls, internal walls, external covering of external walls, ceilings, roof, etc.). The bill of activities can be used for calendar planning or for the schedule of payments in contract management.

Unit price description of work – includes the description of the work in catalogue form, the unit price together with unit of measurement and the code of the unit price. As each catalogue of unit prices aims to be the most widely used, it must consist of the most common construction technologies. As the first use of unit prices in the cost-control process is for estimates during the design stage, it is logical for the client to use the same data (without prices and costs) in the bills of quantities when the invitation to bid is issued, thus ensuring comparability of bids from different bidders. It is therefore recommended to use a country's most widely used standard catalogue of average prices. Usually in every country there are one or two specialised construction-estimating companies that can provide the catalogues of such unit prices. Where possible, contractors use the same codes in their cost databases and descriptions of works in order to make the bidding process less labour-intensive for them. Of course, prices and costs are different in every contracting company. However, it is a necessary step in standardising IT technology in building that leads us to the automation of data processing and data communication between partners (client, designer and builder).

Unit prices in the catalogue and estimates of costs can be presented as separate components: the cost of materials, the cost of labour and the cost of plant. Unit prices can be presented as direct costs, or costs including average overheads and the profit of contactors.

Specification (description) of works or elements of building – describes quality levels and measures or other conditions of works not given in the unit price description of work, or in the drawings. As in the case of the estimates and the bills of quantities, the specification (description) of works and corresponding quantitative information should be presented in groups according to classification standards of building costs.

Prices of buildings and building elements – average m^2 or m^3 prices for buildings grouped by similarity, where the total price of building is divided in percentage terms between building elements according to standard classification (for example, external walls, roof, internal finishing, internal water supply, etc.). Such a catalogue can consist of about a hundred building groups, with every position illustrated on a scheme drawing in order to aid identification. Extrapolation and interpolation of building volume, proportion of parts and price indexing can be used to find the most precise prices in the early phases of the building's life cycle.

Construction cost estimate (cost plan, budget) – the name (and variations) of the cost calculations at the discretionary phase in the construction life cycle. To ensure comparability between estimates calculated by different consulting companies or bidders at different times, it is necessary to use standardised unit prices, standardised descriptions of buildings and cost classifications, as well as standardised forms of estimates for different phases and different users. As the project develops, the cost estimate will be made more exact and more detailed. Construction cost estimates can be classified according to their depth of cost analysis, phase of management, the norms used and the methods of calculation.

❏ **Detailed unit price estimate** – is calculated for every building and for every facility associated with a building. Detailed estimates include all elements of materials, labour and plant. The estimate is called either the estimate of unit price works or the estimated cost of resources, depending on whether the estimate includes the cost of all work resources together (required for the work), or estimates them separately. The detailed unit price estimate is presented in the form of a table, with the costs grouped, and contains sums of the construction costs either grouped by building element, or in the form of aggregated works following the national standard of classifying building costs. The detailed estimate is therefore the basis for the *object estimate*.

❑ **Object estimate** – cost calculation composed of the results of unit price estimates for every building, structure and facility of the developing project. The object estimate is the table heading on which is the name of an object (building or facility). The rows of the table are names of building elements or aggregated works, according to the standard classification of construction cost. The names of the columns (if such division is necessary) are the elements of construction cost, i.e. the cost of material, labour and plant.

❑ **Master budget (master estimate, master plan)** – is composed from results of the object estimate and includes all expenditure connected with investment in the project. The document is often set out in the form of a table. In the rows of the table, costs are divided according to the standard classification of building costs (costs to the owner for the acquisition and freeing of the land, the costs for design, management and supervision of the building, the cost of construction and the cost of preparing the building's use, etc.). The columns list these costs separately for the buildings, structures and facilities of the project by producing object estimates. The master budget is a document for the owner and the financiers.

Methods of cost planning (estimating) – the choice of method will depend on the objectives, detail of data and the obligatory norms relating to public projects:

❑ **Functional analogy method** – presents the aggregated unit prices of functionally similar buildings. These can be units characterising the functional use of the building (for example, bed unit for hospital buildings, pupil unit for schools or living m^2 for dwellings, etc.) or m^2/m^3 as parameters of a building-space programme. This method is used in the profitability-analysis phase when structural decisions about the building are undetermined.

❑ **Structural analogy method** – uses aggregated unit prices from structurally similar buildings. Principal structural decisions are made, and the main building measures are usually determined in the phase of scheme or preliminary design. This gives the possibility to express the total unit price of a building as the percentage divided between the structural elements (such as the basement, external walls, internal walls, roof, etc.) and the technical systems operating inside, or in association with, the building (such as water supply, ventilation, heating, etc.). This method enables the possibility of evaluating the cost of resources more exactly, as it can take into account the technical differences between the actual building and the building analogy and parts thereof.

❑ **Resource estimation method** – used in the basic design phase and in bidding calculations, when a building and its elements are dimensioned, the bills of quantities produced and work and material quality parameters are set. In the design phase, if the building contractor is not yet known, the national or regional average unit prices of materials, labour and plant are used. In building companies the cost prices or bidding prices are used. Often in a single estimate, the prices of works and prices of resources are used. As a result of resource-method estimation, we obtain a price based on direct costs, to which the overhead cost and profit will be added in the tendering process to arrive at the tender price.

Contract's price mechanism – describes the alternative methods by which partners can manage the pricing process, depending on their preference for dividing financial risks between themselves. Put simply, pricing mechanisms can be divided into two groups:

❑ **fixed price** agreed before signing the contract;

❑ **cost reimbursement** of all the reasonable costs that become clear after the completion of work.

The choice between the two must be made during the very early stages of project development. The following is a description of **fixed price** approaches:

❑ **Fixed, lump sum price with added bills of quantities**: lump sum is fixed, but the bills of quantities and specification of works (descriptions of quality) are included in the tender documents. The owner procures the bills of quantities and control-estimates during the design process. This method makes it easy for a client to compare bids from different applicants, to choose the successful applicant, negotiate the price and deal with changes during the development of the project.

❑ **Fixed, lump sum price with bills of quantities prepared by building contractor** – leaving the measurements of works to the contractor makes it difficult to compare different bids, despite the existence of works' measurement standards. Although it increases the financial risk of the owner, this method can be used to speed up the process.

❑ **'Tailor made' price** – can be used in contracts with a 'trusted' contractor, who will be involved in a project in the early stage of design. The contractor's experience can be used to find technically and technologically economic solutions.

❑ **Price on schedules (price list)** – used in contracts where only the prices are fixed, not the amount of work to be done. Used in contracts for routine and repetitive works (repair work on streets, for example).

Cost reimbursement contracts – are used when it is difficult or impossible to measure amounts of work, and where the nature of the work is unclear, or when very high-quality work is required and must start immediately, for example, after an accident, etc. The fee in such cases will be added to the reasonable and actual prime costs of labour, materials, plant and temporary works. The fee can be contractually agreed in three different ways:

❑ as a **percentage** of prime cost;

❑ as a **fixed mark-up** agreed within the contract;

❑ by calculating the approximate fixed sum and including it in the contract as a **target cost** and agreeing on the norms affecting the fee, depending on the relationship between target and actual cost.

The cost-reimbursement method requires that the client controls the contractor's cost documents and means that the financial risk for the owner is higher than when using fixed-price contracts.

Ceiling price of construction – a limit of construction costs for the designer. The price determined at the profitability-studies phase and used in financial plans. A ceiling price can be used in the contract with the designer as the upper financial limit, stipulating that the designer find more economical solutions for the project.

QSC calculations – the bills of quantities (Q), the specifications and descriptions of work (S) and the unit-price or resource-based cost estimates (C) form the basis for balancing the owner's targets on the cost, quality and the duration of the project. These data are closely linked from the point of view of design and construction-management data-processing tech-nology. Dealing with these links as a network results in sub-stantial savings in the costs for the design and construction project management. It is necessary here to pay attention to the complexity of these calculations, because owners trying to save money in the early phases of a project seldom request these results from consultancy (design) companies with the related links between them. The links are necessary in the later stages of procurement, construction management and building main-tenance. This network of links is formed in QSC calculations but is not fixed for future use.

The first step of QSC calculations is to measure the quantities using CAD programs, a standard of quantity measurement and a digitiser. The result is the bills of quantities.

In the second step, it is necessary to describe the quality requirementsof materials, building elements and works to the level of quality standards of resources. The result is the descrip-tion of works or specification.

The third step is the calculation (estimate) of the cost of works and materials. Using IT enables the detailed unit-price estimate to be produced without additional input, as during the second step (the descriptions of work) the codes of prices are chosen – the rest is automated work carried out by computer, based on standard algorithms and methods of estimating. The result is a detailed unit-price estimate or resource-based cost estimate.

From the above description, it is clear that from the owner/client's point of view the expense of both time and money is minimised if QSC calculations are obtained in a more complex form from the design company, rather than leaving the measurement of quantities and choice of quality requirements to the construction contractor. Detailed bills of quantities at unit-price level can be used by evaluating the price bids from tendering builders. The costs for QSC calculations make up about 0.5% of overall construction costs; however, they substantially decrease the amount of cost risk to the owner. If the owner does not pay the necessary attention to QSC calculations, then ultimately the results of the project often do not correspond to the owner's expectations. In this manual special attention is paid to QSC calculations because the general conditions of contract (between the owner and the designer, and between the owner and the contractor) and standards of design do not cover the subject. QSC calculations are one of the main requirements for achieving the owner's project objectives during the building stage.

Chapter 1
Preparation stage

Manual of Construction Project Management for Owners and Clients, First Edition. Jüri Sutt.
© 2011 John Wiley & Sons, Ltd. Published 2011 by John Wiley & Sons, Ltd.

1.1   Project statement

For an owner the *project* begins by the formulation of goals at the conceptual stage and progresses by the appointment of the project manager. The goal might be to construct a building that the owner will use himself, or a building he will sell, or a building from which he will take revenue from rent. The goal should be set in such a way that measurement of the extent of success can be made. The most important parameters in the process of assessment are: cost, quality and time.

Goal

The goal of the project statement is to determine the scope of the project and its participants, to quantify the goals and the assessment criteria, and to prioritise these goals. Usually the main goal is to minimise the cost. The other goals – quality and time – can be taken into account as constraints, although it is possible to 'weight' each of the three criteria and then assess the weighted rating. In doing this, it might create an opportunity for possible corruption. For example, an unscrupulous administrator could set up the formula for weighted criteria so that his favourite contractor can take advantage.

Activities

❑ The initial quantitative determination of the goal using the building's units of power or its volume, or its available area. It is likely that there is a need to specify the goal more than once during the project's development, considering the financial, time and quality demands and constraints.

❑ Prioritise the goal's attainment criteria for both cost and time. Clarifying the relationship between the goal and the financial constraints is also a responsibility of the relevant consultants, advisers and professional project managers during project development. The early identification of priorities in this regard enables the most suitable procurement scheme to be chosen. To select the procurement scheme, consideration of the following alternatives is necessary in order to make the correct choices for the project:

1) **Time** goals:

- possible short duration for the project as a whole;

- possible short construction duration;

- possible early start of construction works;

- exceeding the contract's deadline is extremely undesired;

- completion earlier than the expected deadline is not recommended;

- necessary step-by-step implementation of the project, including identifying the installation time of necessary equipment.

2) **Cost** goals:

- minimum cost for the whole project (land purchase, arrangement of procurement, design, construction);

- the best economic balance between the cost of construction and the cost of the building's usage period;

- minimum capital costs;

- minimum usage costs;

- minimum construction costs;

- exceeding the agreed construction price is extremely undesired;

- achievement of possible early certainty of price ;

- minimum implementation of owner's equity;

- division of risks using contractors as investors.

3) Is the **probability of change** to the project's initial technical and technological data high or low?

4) Does the owner want to **participate actively** in all stages of the *building-procurement* process?

5) **Division of risk** between owner and contractor. It is important to bear this in mind when choosing the procurement scheme.

6) Is *competitive bidding* recommended or necessary?

❑ Choose the general procurement scheme (traditional contract, *design and build* contract, professional-management contract) and *pricing mechanism,* which can be accomplished with the help of the contract's principal adviser.

The outcome

The outcome is a protocol containing fixed objectives for the quantity parameters of constructing the building (m^2 for dwellings, number of bed units for hospitals or km of gas pipeline, etc.), limitations and priorities for cost, quality and time based on choices from the alternatives listed above. A decision as to whether the principal adviser is indispensable or not follows in the next stages of the project.

1.2 Appointment of a project manager and integration of project management in the process management of the owner's company

Goal

The goal here is to appoint a project manager (and his team) because the next steps of the project need special construction-management competence. A decision must be made as to the method of collaboration between the project manager and the owner's management team.

Activities

❑ Appointing an in-house project manager who must be available throughout the project duration and who is the single contact for the organisation. Authority to deal with project issues inside and outside the organisation is delegated to this person.

❑ Depending on the size and the complexity of the project, there might also be a need for a project-management team and special committee with different areas of expertise. The project manager is responsible to the committee.

❑ If it is impossible to find a project manager in-house who has sufficient time and ability, the contract's *principal adviser* should be found externally, and the working relationship will be based on a mandatory contract.

❏ The primary tasks of the principal contract adviser are to contribute by identifying the owner's needs using profitability studies and by preparing the arrangement for procurement. It might prove practical to include the principal contract adviser in project management until the end of the project.

❏ Positioning the project manager/principal contract adviser in the client company's management structure. This is done with the view that the company's functional divisions should serve project-management structures in the preparation of initial data and in the integration of the adopted solutions.

❏ Determining the time limits for the integration procedures.

❏ Carrying out the juridical procedures required to give authorisation to the project's manager/principal contract adviser.

❏ Motivating the project manager to guarantee the owner's priorities and goals.

The outcome

Project-management functions are identified, as are project-management hierarchy in the company and the relationship with the client; the time schedule and the cost plan of the preparation-stage activities are prepared.

1.3   Needs and profitability analysis

Preceding procedures

These analyses are dependent on the company's strategic development plan and the project's preliminary business plan.

The version of the company's strategic development plan and business plan that brings the solution of problems via real-estate purchase, sale or rent is not the subject of this handbook. In what follows, the owner's activities are described sequentially, specifying goals for the solution of real-estate development problems by erecting new buildings or reconstructing existing ones.

Goal

The goal is to find the best solution (profitability or social benefit) for the proposed building-investment project taking into account the cost of construction and cost of usage of the building. The acceptable level for the project-development costs and the proportional split between building cost and the cost of land purchase would be determined. This entails the determination of the *ceiling price* for the building. The first version of the *scheme design* should be produced together with the needs analysis. Drafting the scheme design enables money to be saved while the preliminary design is produced, especially when the same design (consultancy) company is used for both stages of the design and cost calculation. The needs analysis, together with the drawing-up of a corresponding alternative scheme design and profitability calculations, is not a stage during which to make financial savings, considering that the costs of this stage are very small compared to the costs of subsequent stages, although their effect on project efficiency is most significant.

Executor

The executor of the needs analysis is the owner himself, using advisory services provided by the design (consultancy) company. Calculations of project profitability should be carried out by a person or company with competence in the field of a building's life-cycle economy. It will be necessary to consult with construction-estimating companies, bank-loan departments, real-estate companies. Professional construction-management companies, as well as design companies, can offer complete analyses. The exact division of work between the owner and the consultancy company depends on the project type and will be agreed contractually. The most common advice is to procure the scheme design with building-cost calculation based on the functional-analogy method from the consultancy company, with subsequent profitability calculations made by the owner himself. Collaboration between the owner and the consultant must be very close.

Cost and quality assessment

During the phase of needs and profitability analyses the owner's subject for evaluation is the investment project, which involves all the building's life-cycle phases including corresponding **costs** and **revenues**.

A. Building costs (capital cost, single cost):

❑ Costs related to site possession, including the price of the land, financial costs (loan interests), notary fees, state taxes and fees, survey fees, brokerage commissions, evaluation fees, court costs, land utilisation fees, charges for release from ties on property, recompense costs for rent contracts, costs for purchasing rights for the property.

❑ Costs of site preparation and occupation, including the costs for building protection and demolition as well as costs for elimination of threat, preparation of surface, replacement of restrictive utility networks.

❑ Cost of construction.

❑ Cost for furnishing the building.

❑ Cost for the owner's project management, investigations (needs and profitability analyses, feasibility studies), design, the owner's construction supervision, recruiting and training the personnel to work in the prospective building.

❑ Cost for technical equipment, for example, stage machinery in a concert-hall building.

B. **Exploitation costs** (only current costs, which depend on building costs, are listed):

❑ Heat, electricity, water, sewerage, communication costs.

❑ Maintenance costs for the building and its technical systems.

❑ Repair costs for the building and its technical systems.

C. **Exploitation-period revenue**:

❑ Revenue from the investment project.

❑ Profit from the sale of real estate.

❑ Profit from arisings after building demolition.

The determination of the quality requirements in this phase depends on the determination of the space-planning programme (functional quality) and on the calculation of the proportional split between building and exploitation costs. To accomplish this, it is necessary to balance the costs of building elements (non-recurring expense) and the cost of each element's use (current cost) against one another, for instance:

❑ Building costs to ensure thermalinsulation of the periphery structures (walls, roof, windows, etc.) – heating costs.

❑ Building costs for ventilation systems (for example, natural ventilation or a system with different heat-exchange systems) – ventilation and heating costs.

❑ Building costs affected by the life-expectancy of the roof material – roof maintenance and repair costs.

❑ Building costs depending on the quality of the external-envelope materials – external-envelope cleaning and maintenance costs.

The determination of the proportional split between building and exploitation costs depends on the owner's relationship with the building after its completion. Depending on whether he will be an occupant or whether he is building for sale or for rent, or whether he is a public or non-commercial developer, his decisions tend to lead to economies in either the investment or exploitation costs.

Base for evaluating costs

When evaluating building cost in this phase, appraisals per square metre or per cubic metre – based on the functional-analogy or structural-analogy method – are used, with consideration given to the differences between the analogue and the building in development. Widely available cost-estimation software based on building space and room parameters can also be used. If the plot of land already exists, then the building's quantities of external technical networks are measurable and the corresponding costs can be calculated. To evaluate the owner's total project cost in this phase, it is necessary to add the following to the construction cost: 3% for the owner's investigations at the preparation stage, 5–15% for design, 0.5–1.5% for the owner's supervision, 2–6% for the owner's management and 10–20% for the owner's financial cost and costs involved in obtaining the land. Costs for compiling the detailed plan as well as costs for connecting to the technical networks may be added to the above. This evaluation, based on specialised literature, is approximate although it gives an overview of the magnitude of costs.

For evaluation the owner should make price enquiries, use the services of consultants, investment and credit companies, study real-estate brochures and web portals. The error in the evaluation of building costs can be up to 25% in this phase.

Activities

☐ Drawing up the space-planning programme based on the company's development strategy. This is based on technological schemes of production or other main activities, production capacity, reserves of raw materials and end product, number of employees in categories, transportation schemes, requirement for aid and service rooms, the location of businesses within the building and the need for parking within the site and outside the site. In the case of construction for dwellings, this is the point to determine the number of floors, apartments and their structure, as well as ensure comfortable conditions, extra needs of thermal and acoustic insulation etc. While compiling the space-planning programme, designers/consultants assist the owner as they possess empirical-data sets for space-planning or surface-planning needs (square metre or cubic metre per person, etc.).

☐ Functional objectives and chosen principles of dimension are given in memos, tables, (technological) activity schemes, plan briefs and room specifications.

❏ In the case where the development objectives can be met through the *reconstruction* or *renovation* of existing buildings, suitability of the existing *detailed area plan* with the prospective building's purpose of use should be determined.

❏ Producing brief drawings of the building and the plot's general layout with the aim of establishing dimensions. The listed drawings, together with the technical and economic characteristics, are retained with any explanatory notes about the initial amendments to the scheme design made during the phase of profitability calculation.

❏ Determining environmental requirements.

❏ Determining quality standards of rooms and plots (fire-safety category, heating, lightning, ventilation requirements, etc.). Quality requirements are designated as functional needs, not as specified solutions.

❏ While determining the quality standards, the effects of quality on cost should be considered. In the preparation phase it is useful to use the above-mentioned estimating software that employs the space-planning programme and functional-quality characteristics of rooms and facilities.

❏ Determining the transport and traffic schemes, and the plot's maintenance requirements.

❏ In the case of an industrial building, heat, water, gas, electricity and communication needs should be determined. Assessing the amounts of foul water and any harmful manufacturing waste (liquid, solid, gas) and the methods for disposal (or possible recycling). Initial environmental-protection calculations are made if necessary. For instance, obtaining normative levels of gas waste depends on the height of the chimney and its distance from neighbouring buildings.

❑ In addition to the buildings, engineering facilities are listed on the site plan along with an explanation of their main technical characteristics. These form the basis from which to evaluate building costs in alternative solutions during the phase of profitability analysis.

❑ Determining the possible structural types of load-bearing elements and enclosures (roofs, walls, etc.), as well as possible heating and ventilation system alternatives, if the owner has his own such preferences.

❑ Preparing the list of necessary alternative solutions. Checking whether all the necessary initial technical characteristics for the project's economic calculations have been established.

❑ Evaluating the *project's* costs for all alternative solutions. It should be borne in mind that in this phase the evaluations should be made in both absolute values, in order to decide on the expediency of investment, and in relative values, in order to make a choice between the project's alternatives.

❑ Choice of preferred profitability-calculation method. Two methods can be used (see Table 1):

• Firstly, the **net present value or cash flow method**, in which, on the basis of the future accounting exploitation period's revenue and cost, net profit is calculated by years and discounted to the year of investment in order to make it comparable with the capital (single) cost of the investment project. Using this method, the choice of

Table 1:   An example of the effect of the choice of assessment method and length of accounting period on the choice of best investment project option

	option 1	option 2
Investment cost	20.0 mil €	15.0 mil €
Heating cost	0.5 mil €/year	0.9 mil €/year
Maintenance cost	0.6 mil €/year	0.6 mil €/year
Revenue from rent	3.6 mil €/year	3.6 mil €/year
Net profit	2.5 mil €/year	2.1 mil €/year
A. Internal rate of return		
1) Assessment criterion: payback time	8 years	7.1 years
B. Net present value		
Assessment criterion: profit during assessment period		
Duration of accounting period		
2) 10 years	2.5 × 10 – 20 = 5 mil €	2.1 × 10 – 15 = 6 mil €
3) 20 years	2.5 × 20 – 20 = 30 mil €	2.1 × 20 – 15 = 27 mil €
4) 50 years	2.5 × 50 – 20 = 105 mil €	2.1 × 50 – 15 = 90 mil €

Remarks:
1) the choice of the best option from the two alternatives depends on the choice of criteria from 1 to 4 in the calculation above (best options shown in grey);
2) for simplification, the influence of possible changes to inflation and profit margins are ignored.

accounting-period length is important, and this decision must be made by the owner himself. In cases where a shorter period is chosen, the owner is opting for a higher investment cost, whereas in cases where a longer period is chosen, the emphasis falls on exploitation costs and revenue. This will, in turn, depend on the reason for construction: occupation, rent or sale. At this point it should be said that the economic accounting period does not coincide with the potential exploitation period. The exploitation period of a building is usually about 50 years. At the same time the building's structural stability, which is usually guaranteed, is not the foremost consideration but rather its function (target of use) and the comfort requirements that are linked to this function.

- Secondly, the **internal rate of return method**, which is generally used when assessing speculative business projects. The payback time in years is found as the ratio of investment costs to the project's annual net profit. For example, there follows an illustration of the importance of time as a factor when making economic assessments of investment projects, or their alternatives.

❑ The owner's required start and finish dates for the project are fixed, including corresponding dates for design and construction.

❑ The profitability analyses of the options and of scheme design can lead to changes in the company's development strategy, including change in the initial conditions for the building and new technical and economic calculations. Certainly the scheme design and cost evaluations in this phase are approximate, although in the case of an industrial building only this method makes it possible to determine the minimum land size, with its presumption to solve the accompanying infrastructure problems.

❑ The profitability study ends either with the acceptance of the investment decision, or with the decision to stop the project.

❑ The approval of the scheme design in the profitability-study phase is given if the decision is positive.

The outcome

The results of this phase are fixed functional objectives, the space-planning programme, scheme drawings, the calculation of the length of the building's accounting exploitation period, profitability calculations and the investment decision.

Activities directly following the investment decision

❑ Seeking an appropriate *plot of land*.

❑ Requesting design conditions from the local authority.

❑ In cases where the land already exists, feasibility investigations should be carried out (probably not as many as for a new plot).

❑ In the absence of a *detailed plan* in the area where, according to the general area plan, it is necessary, or if the conditions shown on the detailed area plan are unsuitable, then the detailed plan should be created.

❑ Choosing a financial scheme.

1.4   Preparation of the financial scheme and loan contracts

Preceding procedures

Profitability analyses, and based on these, acceptance of the investment decision.

Goal

The goal is the planning of the cash flow and preparation of loan agreements for the project. It should be taken into consideration that in the comparative analysis of options the dynamic represented by financial conditions (inflation, etc.) is usually ignored because its effect on all the options is the same; however, this is an essential factor while preparing the financial scheme. The choice of the financial scheme depends primarily on the owner's relationship with the building:

- ❑ The owner is the occupant – he intends to exploit the building in the long term, meaning that he will need a long-term loan.

- ❑ The owner is building for sale, which means that he needs a short-term loan and/or he involves the buyer's and/or the builder's finances.

- ❑ The owner intends to take a profit from the investment using rental income, requiring a long-term loan and/or involving the builder's finances.

- ❑ The owner is a government, or a public institution or state is guarantor of the loan, in which case the conditions of financing are regulated by the state.

The owner considers the future costs of the project's various stages in two ways. Firstly, the costs involved in purchasing the plot, building design and building erection can be described as 'cheap money' for the owner as it is borrowed, with payment coming from future project revenue. Secondly, the costs at the project-preparation stage, despite being lower in order of magnitude, are financed from the company's current account and can therefore be described as 'expensive money' for the owner. In hypothetical terms: saving 1% of this 'expensive money' from the gross value of the project might result in the building becoming 10% more expensive, or there might be some loss in the project's functionality, something that is not always easy to measure in financial terms.

Executor

Although solving the financial problems ought to be the responsibility of the financial specialists of the owner, rather than being one of the construction client's responsibilities, the investment's building character dictates various specific considerations about the economics of the building's life cycle, which necessitates the involvement of a specialist in building economics (the quantity surveyor).

Activities

❏ Preparing the initial calendar plan of the project and its approval.

❏ Determining financial schemes and potential sources: the owner's own resources, loans, the resources of the builders and the building's occupants, share sale, possible subsidies, etc.

❏ Forecasting the project's rate of profit, rate of loan interest, buildings' cost index, inflation rate, rent and sales prices during the period of the financial plan. Identifying the amounts of the necessary fees.

❏ Producing the financial plan. One month is recommended as the unit of time interval to guarantee the necessary accuracy of discounting following the phases of project development.

❏ Determining borrowing objectives: the costs of the project-preparation period preceding design, purchase costs of the plot of land, design and building costs. It should be remembered that building costs do not usually cover all possible costs, for instance, the cost of purchase, installation and commissioning of technical equipment needed for production in the erected building.

❏ Determining the length and stages of the loan period (short-term and long-term loans).

❑ If it is possible to divide the building's implementation into sequences, then it should be calculated in positive (profit) as well as in negative (cost) cash flow. It is important to include the division of the implementation of the building into stages in the initial task for design.

❑ The financial plan should include steps (project-development time milestones) to specify the project's costs and the costs of corresponding cost calculations. From the viewpoint of the building's project development, the following milestones would be typical:

- completion of feasibility studies and corresponding phase of scheme design, during which the *ceiling price* of the building is determined, from which the designer should work;

- approximately 30% readiness of the *preliminary design* (general solutions are chosen), when making changes in the project (design) with the aim for correct, controlled construction costs that do not involve a large increase in design costs;

- 100% completion of the *basic design*;

- building contractors are chosen and contracts are signed;

- final settlements are made with the building contractors.

❑ Preparing the short-term loan conditions relating to the reimbursement of the feasibility-studies and multiple-option scheme design expenditure. A corporative or project-based loan could be used – and it might be practical to repay that loan from the project-based long-term loan.

❑ Preparing the long-term loan conditions.

❑ Considering the possibility of a mortgage for the loan.

❑ Evaluating project risks outlined during the project-statement phase (see Section 1.1). Based on experience, there should be a reserve in the financial plan of not less than 10%.

❑ Preparing the financial plan in the form of charts and tables.

The outcome

Project calendar plan, cash-flow plans, financial sources, financial scheme for project financing, mortgage-appraisal documents, financial contracts.

Activities that follow directly after the formulation of the financial plan and preparation of the loan conditions

Signing the loan contracts, land purchase.

1.5   Preparation for land purchase and corresponding contracts

Preceding procedures

The investment decision is approved. The requirements resulting from the needs and profitability-analysis phase are determined: the size and shape of the plot of land, distance from the transportation network and any financial restrictions applying to the purchase price. Searching for a suitable plot of land should begin before the end of the profitability analyses.

Goal

The goal is to obtain a suitable plot of land or acquire the leasehold of such a *plot of land*. Apart from this main goal, the following should also be taken into account:

❏ Purchase of the land could be a profitable long-term investment separate from the building-investment project.

❏ The land should be saleable in a short period if, during project development, it turns out that construction on the plot is impossible because of a conflict arising from the design requirements, owing to environmental reasons or the neighbours' interests, etc.

❏ The need for possible enlargement of the company activities in the future.

Executor

The executor of the land purchase is the owner himself, or his principal adviser or the professional construction manager. Using the help of real-estate companies is not recommended as a conflict of interest could arise as they are the biggest land-owners and their recommendations could have self-seeking interests.

Cost and quality assessment

The land's value depends on:

❏ Distance from the transportation network (harbour, railway station, etc.) or from raw-material sources.

❏ Existence of facility networks (incoming services, etc.).

❏ Binding prescription of land use according to the approved detailed area plan and the possibility of its change arising from the higher-level area plan.

❏ Restrictions for building, i.e. the number of structures on the plot, determined by the detailed area plan.

❑ Limits on building height, maximum size of the building area of the plot, etc.

❑ Aesthetic considerations.

❑ Neighbourhood.

Depending on the particular investment project, the importance of these factors will vary. The most important factor might be the distance from the catchment area or centre of attraction.

Prices of suitable plots vary widely and depend on supply and demand, in other words, on general investment trends. To obtain an overview of current market prices the owner should compile a list of the sites available. This should be done during the profitability analyses, prioritising sites according to the value characteristics that are important for the project. At the same time the owner should determine the maximum price he can afford to pay. To do this, the design, building, *building-management and supervision* costs should be deducted from the investment costs calculated in the profitability analyses. In cases where the project is an industrial building, alternative calculations should be made taking into account the end-product and raw-material transportation costs, the latter depending on plot location. At the same time variations in salary should be considered – it may be that the more distant the site is from a local centre, the lower is the salary.

Activities

❑ Advertising the desire to purchase in real-estate brochures and web portals.

❑ Analysing the sales advertisements placed in real-estate brochures and web portals.

❑ Dealing with enquiries from real-estate companies.

❑ Consulting with local-government land experts.

❑ Inspecting suitable plots of land; negotiating with landlords.

❑ Carrying out feasibility studies (see Section 1.6). Leaving aside some steps in the feasibility studies prior to land purchase increases the owner's risks significantly, and, as the time saving is less than one month, it is inadvisable.

❑ Making the decision on the purchase of the land.

❑ Requesting purchase permission from local government. Even if the law does not require this permission, it is useful to obtain it anyway in order to reduce the owner's risk, especially when dealing with a larger project and there is a need to change the purpose of land use as specified in the detailed area plan. In some circumstances it is necessary to request permission from the authorities, for instance, when the plot is agricultural land and the purchaser's aim is to use it for a purpose other than agriculture, or when the purchaser is from a different country.

❑ If the purchasable real estate needs to be divided, then geodetic measurements should be made and a site plan drawn up, and both should be submitted to the local government for approval.

❑ If there is a need to produce a detailed area plan, then an agreement of cost distribution between the buyer and the seller should be made.

❑ Drawing-up the *servitude contracts*. Servitudes could be necessary for the land in favour of neighbouring plots, or vice versa.

❑ Preparing the purchase–sale contract together with an attached plan of the land. If there are any doubts concerning the project's development result (detailed area plan approval and change of land use, acquisition of real estate, etc.), then the contract should contain warranties and other conditions to decrease risk.

❑ Determining the division of the transaction cost (notary fee, state fee, etc.).

❑ The notary should prepare the contract, which should then be signed.

❑ Registering the acquisition of the real estate.

The outcome

Comparison charts of suitable plots of land with criteria assessments, details of plot inspections, minutes of the meetings in which these decisions were made, preliminary contract, application for purchase (if required), plans and contracts of any servitudes, purchase–sale contract, real-estate registration documents.

Activities that follow directly after making the land-purchase decision: completion of feasibility studies.

1.6   Feasibility studies and corresponding investigations

Preceding procedures

A plot of land satisfying the conditions of the development project is found.

Goal

Finding the natural, technical, juridical and economic conditions for use of the proposed land. If there are buildings or other objects on the land, the investigations include these as well.

Executor

The executor is the owner himself, with the help of his adviser. Using the adviser at this point makes it possible to substantially reduce the time taken. Advice can be obtained from the principal procurement adviser, from the professional construction-management company and from the design company.

The basis for cost evaluation

The initial data for estimating construction costs in this phase come from the plan of the land, the schematic plans and sections of the building from the profitability-studies phase. From this base it is possible to establish rough measures for the infrastructure facilities and identify the area and volume of the building. The plot plan will be sketched with roads, courtyards, technical facilities, water supply and rainwater disposal, waste disposal, landscaping, fences and communications. For industrial buildings the initial data include the technological schemes of production, capacities of raw materials, ready production and waste, the dimensions of production lines, the height and fire-resistance of rooms, schemes of ventilation and lighting, etc. Often the owner already has some preferences about building materials in this phase, for example, the material to be used for floors or external coverings. The scheme design of the feasibility-study phase is at the same time a suitable initial base from which to derive the detailed area plan and the detailed and complex design; it can be used as an appendix to corresponding contracts. Norms for cost estimates relating to this phase are the same as those relating to the profitability-study phase. Calculations will be amended based on specified measurements, the layout of rooms and other specific needs for the conditions of the rooms. Additional data relating to any external facilities on the defined plot of land make it possible to decrease the estimating error from 25%, during the profitability-study phase, to 15%. The cost appraisal of this phase can be used as the specified *ceiling price* for the designer.

Duration of the phase

The duration of activities is at least 1 to 4 weeks, using an optimistic appraisal. This optimistic appraisal depends on combining activities from a timescale point of view. This can be argued for on business projects with high profit levels. Arranging activities so that they follow one after another substantially prolongs the duration of the whole project. However, performing these activities simultaneously does increase the risk to the owner.

Shortening the duration of this phase by cancellation or postponement after land purchase also increases the risk to the owner and likewise increases the risk of error in cost appraisal. The detail level of the feasibility study establishes the presumption to ensure the minimum total duration of design and construction.

Activities

❑ Negotiating with local government with the aim of obtaining preliminary approval for the building-development plan. A negative position within the local authority or community hints at the possible need to find a new plot of land.

❑ Becoming familiar with the community's or town's general area plan and the detailed area plan of the plot, with the aim of understanding all restrictions and conditions on land use, as well as those applying to the buildings on it. Clarifying from the general area-development plan: traffic schemes, water and electricity supply, environmental (natural and cultural) restrictions, etc.

❑ In the absence of the detailed area plan for the land concerned, in a district where, according to the general area plan, one is necessary, becoming familiar with local practice in the drawing-up and agreement of the detailed area plan.

❑ Becoming familiar with cadastral, real-estate and building-register data, as well as with the documents that approve ownership of real-estate and rental contracts.

❑ Performing actions arising from the necessity of dividing, measuring and registering the acquisition of real estate.

❑ Performing actions relating to the control obligations covering property using the real-estate register, real-estate mortgages, contracts of servitude, rent contracts, etc.

❑ Becoming familiar with any plans submitted by neighbours and owners of infrastructure facilities. If necessary, obtaining relevant agreements from neighbours.

❑ Becoming familiar with the cadastral register in order to establish whether any restrictions are applied to the holding of this plot.

❑ Investigating geological conditions, such as ground-bearing capacity and hydrological conditions (for example, the level of ground and aquifer water, water resources), using archive data from the relevant authorities. If necessary, carrying out additional investigations.

❑ Performing preliminary assessments of environmental impact.

❑ Becoming familiar with any restrictions in relation to protection of nature, culture or antiquity.

❑ Becoming familiar with conditions for foul and surface-water drainage (including the use of gravitation for drainage) and requirements on treatment of foul water.

❑ Becoming familiar with access conditions onto local main roads (using the local authority's general area plan of community development and the road authority's development plans).

❑ If the project involves hazardous substances such as dangerous gases, special preliminary calculations must be made taking into account the necessary safe distance from neighbouring buildings.

❑ Examining existing buildings and facilities on the site.

❑ Establishing the conditions under which the project interfaces with existing infrastructure systems (technical and economical), requesting the official documents to enable connections to this infrastructure (authorities usually require that their own printed forms are used).

❑ Finding geodetic and topographical plans of the plot. If these do not exist, it is necessary to requisition measurements while the land-purchase contract is being prepared.

❏ Inspecting the nature of the plot in order to determine the conditions that will affect construction. It is necessary to inspect the plot to assess it as a building site because the cost of temporary works and facilities make up 1–12% of building cost, depending on the site conditions and the type of structure. For dwellings and office buildings in built-up areas, it is usually 3–5%. Site conditions also influence the duration of construction, which will have an impact on construction cost and revenue to the owner. If a shortening in construction duration is planned by operating three shifts, the construction costs willrise by 5–6%. If the project is a commercial one with a profit margin of about 10%, then the owner's profit from shortening the duration of construction (by an earlier start date to receive profit and decrease financial costs) will increase by double the amount of the costs payable to the builder. Factors influencing building-site costs are: the size of site and its access; whether existing buildings and facilities must be replaced or demolished; buildings, roads and any adjacent land features such as trees requiring special protection; restrictions on the height of cranes; special restrictions on working in the evenings and at night, etc.

❏ Controlling the influence of site location on the project's programme requirements (start and finish dates, duration).

❏ Assessing the results of the activities listed above and deciding if they give reason to buy the chosen plot of land.

❏ The scheme design of the profitability-study phase must be detailed and the cost estimate and calendar plan of the project must be updated to take account of results at this stage.

❏ Approving the *ceiling price* and the duration of the project as input data for the next steps in the project development.

The outcome

Assessment of the plot of land. If there are existing buildings and facilities on the plot, the examination of their technical condition. Decision on the suitability of the land for construction based on an understanding of all relevant documents connected with the plot, or on the restrictions of use revealed during a site inspection. More detailed scheme design (than in the profitability-study phase). The calendar plan and cost estimate (based on the functional or structural analogies) taking into account the condition of the plot, as well as logistical and infrastructure costs.

Activities following the feasibility study

❏ Decision to purchase the land.

❏ Preparation of the land-purchase contract.

❏ Application to local government to start detailed planning.

❏ Application to obtain technical specifications from local government for design.

❏ Application to obtain technical requirements to enable connections to infrastructure facilities (electricity, water, gas, sewerage, roads, railway, etc.). By law the owners of infrastructure facilities have one month to respond. To find acceptable conditions may need much more time.

❏ Development of the plan of necessary servitudes.

❑ Development of the scheme of procurement and priorities first planned during the phase of determining the scope of the project, using the help of professional advisers.

1.7   Detailed area planning

Goal

The goal of the detailed area plan is to set up through local government – in the towns, villages and other areas that necessitate the use of such a detailed area plan – the land usage and building conditions. It might be necessary to create the detailed area plan if any dividing or joining of land accompanies the registered land purchase, or if there is a need to change the land use from its purpose as designated in the detailed area plan. The cost of the land can therefore depend on whether a detailed area plan has been created, and on the land use as designated in the detailed area plan.

Executor

Local government will arrange the preparation of the detailed area plan, sometimes by assigning a third party. The draft detailed area plan can be drawn up by an architect. Usually the real-estate owner requisitions the work from a design company. In order to save time and money, it is possible to receive the detailed plan from the same design company who assisted the owner in preparing the profitability and feasibility studies and the needs analysis. Several groups are involved in the processing of the detailed area plan: the owners of real estate within the planning area, local residents and other interested parties.

The cost of **producing** the detailed area plan depends to a great extent on the specific project, the site's characteristics, environmental conditions and the need to evaluate the environmental effect. The current cost is approximately one thousand Euros per hectare of land.

The timescale when **producing** the detailed area plan is divided into two phases. Firstly, the formulation of the outline detailed area plan, which could take 1–3 months. Secondly, the process of coordination and approval of the detailed area plan leading to its implementation, which could take 3–6 months; however, if there are conflicts then it is hard to predict the time required. If the owner foresees that implementation of the detailed area plan could take significantly longer than other pre-construction stages, then it is possible to start design before implementation of the detailed area plan in order to shorten the project's overall duration. However, this happens at the owner's risk – the owner cannot apply for building permission from the local government before the detailed area plan has been approved.

Activities

❑ Applying to local government to produce the detailed area plan. Waiting for the local government's decision to take action.

❑ Signing the agreement with local government concerning the composition of the detailed area plan and delegating this to the design company responsible for the detailed area plan.

❑ Compiling the initial data required for detailed planning. The scheme design made during the feasibility-study phase (see Section 1.6) should comply with these initial data. This is the time to address the owner's requests concerning any division of land into smaller plots, as well as road access and traffic regulations, planting of trees and greenery, the requiredsurface area of the buildings, the shape and height of the buildings and their required number, the allocation of facility networks, the purpose of the plot's use, potential threats to the environment (harmful waste and raw materials, etc.), variation from the general area plan, required servitudes, etc.

❑ The agreements and the tasks required to be carried out will be listed in the initial conditions from the local government, such as the required number of plan documents (including whether they should be submitted digitally), the required scales of the drawings, etc. If at this stage there is the need for a preliminary evaluation of the environmental effect, then it would be advisable to follow the same procedure as that used in the project's final phase, in which the application for building permission will be made. It is obvious that during design development most of the parameters will become more specific, although repetition of procedures is normal and saves the owner time and cost in subsequent stages.

❑ Finding the appropriate design company. Negotiations and contract-signing take place through the same procedures used during the construction-design phase (see Chapter 3).

❑ Accepting the draft detailed area plan. Owner's examination and approval.

❑ Submitting the detailed area plan to local government for treatment. Local government then follows subsequent procedures in the order listed below, which means an unavoidable waiting period for the owner. Timescales (enforced by law) are given in brackets and may vary from country to country:

- decision to start processing the plan (1 week);

- notice of detailed-plan initiation published in the press (1 week);

- public display of the plan (2 weeks);

- notification of public debate (1 week);

- public debate and action on the results (2 weeks);

- amendments to the detailed plan and, if necessary, resolving conflicts with local government (4 weeks);

- acceptance and implementation of the detailed plan (2 weeks).

It should be emphasised here that the points above represent only those procedures that cause unavoidable waiting time for the client (arising from legal requirements). They do not include the whole list of procedures that may occur during detailed area plan development.

From experience it is possible to conclude that, on average, the duration of the detailed plan's implementation from start to finish will seldom be less than six months and, in cases where there are disputes, it could be several years.

The outcome

Application to initiate the detailed plan, contract with local government, requirements of local government for detailed plan, owner's initial data for the detailed area plan, contract with the design company to prepare the detailed area plan, submission of plan for public debate, amended and approved detailed area plan.

1.8   Scheme design

Preceding procedures

As building design (Chapter 3) is completed in stages (*prelimi-nary design, basic design and working drawings*), the project-development preparation process (Chapter 1) is also divided into stages, as described in Sections 1.1 to 1.7. During the design phase, progress to the next stage is possible only when the earlier stages are approved, and the same is true of the preparation stage of project development. As the building project (design) is the prescriptive model of a building, so scheme design is the model for a building's development project. If we have agreed that the notion *project* in the building context has two different meanings in several languages, then the notion *scheme design* can also be used in two different ways: the first is how designers understand this notion, and the second is how the owner understands it. The latter entails a complexity of documents produced from the results of the owner's activities during the preparation stage of project development, collated systematically to ensure effective appraisal of the project using three criteria: cost, quality and time. The phrase 'owner's activities' is used despite the input of external specialists, since the owner is the only decision-maker at this stage. This entails a summary to determine the needs and possibilities, the approval of which gives a premise with which to go forward to the next stage: that of building design. It is obvious that there will not be enough architectural drafts to obtain approval for the development project, and, in addition, there must be technological solutions (buildings are not just dwellings), a project cost evaluation, an initial calendar plan setting out dates for the signing of financial agreements, local government agreements about land use, a preliminary *procurement scheme*, planned *price mechanisms* and much more, as described above. Building-design schemes form only one part of scheme design. Scheme design is the outcome of the owner's entire work. It is possible to reduce costs at the preliminary-design stage if in the early stage of preparation the owner involves designers capable of integrating technological production schemes with building and site drafts, and thus offering alternative solutions using CAD programs. Later in the project, producing alternative solutions becomes more expensive and their effect becomes smaller. For this reason it is necessary to have a comprehensive definition of the owner's possible activities and a detailed plan of their outcome.

For simpler projects it is not essential that all constituent activities are detailed in writing or drawn up using CAD programs. The important factor here is that the owner has a precise idea of his possible activities and reaches the relevant decisions.

Goal

The task of scheme design is to set down step by step the owner's ideas, making the overall project's statement (goals, scope, etc.) more detailed, and developing it during the needs analyses, profitability and feasibility studies, preparing the purchase of land, formulating the project's financial plan and the property's detailed area plan. Thus, scheme design can be defined in two stages: scheme design based on the profitability-studies phase, and scheme design based on the feasibility-studies phase. In practice the results of the named phases must not be compiled separately in the named folders. The desired outcome is recording the list of completed decisions of the phase. Separate folders are recommended if external consultants are used. Questions and answers relating to the listed phases are presented in Sections 1.3 and 1.6. For maximum efficiency it is possible to use scheme design as an integral part of the initial task for the detailed area plan (if the project requires an area plan) and for the *preliminary design*. In scheme design, the following should be formally presented:

❑ The project's general characteristics.

❑ Ground plan of the building plot (site) with engineering networks and floor plans and sectional sketches of the building's structure.

❑ Sketches of production technology, building functions and the movement and transportation schemes that are linked to the site and floor plans.

❑ Architectural characteristics.

❑ Characteristics of the building structure, if the owner has specific requirements for such characteristics.

❑ Preliminary data about building geology, the site's geodetic and topographical plans.

❑ Requirements for the following (necessary to formulate for the initial task of design): water supply and sewerage, heating, waste management, electricity and communications installations, fire protection and sprinkler systems.

❑ Schedules of professionals occupying the building, functional requirements of rooms, welfare facilities and ancillary buildings, technological schemes for which the building will be erected, etc.

The client does not have the required knowledge to formulate or answer all the above questions. For that reason it is useful to procure the scheme design from the design company, and involve them during the needs and feasibility analyses (described above). During the first options of scheme design this enables contemporary design technologies (CAD methods, etc.) to be implemented, providing an assurance that different elements of the project are congruent and also reducing work-load, time and cost in the next stages.

Costs in the scheme-design phase, especially in the phase of the profitability studies, are relatively low for the owner com-pared to the costs in the next stages (design and building). However, the effect of the decisions on overall project effective-ness is most significant, and therefore it is not prudent to abandon the multi-option approach during scheme design in order to reduce costs.

Activities

❑ When using external consultancy services, it is useful to follow the procedures described in Chapter 3 for this phase, as it involves the search for the design company and man-agement of design work. If all design work is procured from the same company, a *framework contract* should be consid-ered, which can later be detailed with contracts for every phase, signed after the owner's approval of the results of the preceding phase.

❑ Space-planning programme development, together with rooms' fire-safety and quality-standard classification; deter-mination of room exploitation (number of personnel, utili-sation intensity of the rooms, utilisation requirements) and maintenance objectives (heating, water, electricity, cleaning, maintenance of external areas, etc.), as well as a description of ongoing preventative and regular repairs – maintenance that ensures the preservation of the original quality.

❑ Agreement upon the number of drawings copies to be handed over (including CDs).

❑ Agreement upon the structure of the estimate (budget) for the current stage.

❑ Ascertaining the necessity of a preliminary evaluation of environmental impact; subsequent evaluation and, if necessary, acting upon the results.

❑ Specifying the calendar plan. In the general calendar plan the waiting periods for official permissions are outlined. The method of procurement, the limits on the start dates of design, construction and occupancy are taken into account, along with amendments, debates, etc. The importance of the time criterion and the division of the risk between cost and time should also be considered, as should the effects of seasons.

❑ Approving the calendar plan.

❑ Adjusting and approving the cost-breakdown budget. The budget is formulated using the *structural-analogy method* and is categorised into subdivisions according to the building's cost-classification standard.

❑ Determining the *ceiling price* of the building; the general designer should consider the ceiling price and the choice of the scheme to achieve it (and the motivation scheme); these may be imposed through the design contracts (see Section 3.3).

❑ In the case of *design and build* procurement schemes, determining the project's target cost ('project' in its larger sense), including all costs to the owner. An explanatory note could be added to the *target cost*, listing works and supplies that have not been taken into account but which are necessary for the full exploitation of the building.

❑ Approving the scheme design.

The outcome

The scheme design in the form characterised above comes about as a consequence of the needs, profitability and feasibility analyses and the development of the detailed area plan through systematic and multi-option calculations, using the support of the design company in the early stage of development.

The alternative to scheme design is to perform the calculations and analyses separately, although in this case the data are frequently not interlinked and lack the element of cost calculation. If the initial task for complex design is not produced on such a basis, then project development may not result in the design-stage achievement of the client's specific targets on cost, time and quality.

Chapter 2
Procurement stage

Manual of Construction Project Management for Owners and Clients, First Edition. Jüri Sutt.
© 2011 John Wiley & Sons, Ltd. Published 2011 by John Wiley & Sons, Ltd.

2.1 Choice of procurement path

Preceding procedures

The activities that precede and follow the procurement stage in the client's management-procedures chain depend on the type of procurement method chosen, and therefore it is impossible to list preceding and subsequent activities here. It is recommended that the initial choice of procurement method be made during the preparation phase directly after the determination of the project's scope and goals (see Section 1.1).

The decision as to which procurement path to choose becomes clearer during the project's development process and ends with the choice of contractor and the signing of procurement contracts, contracts for the reduction of risk, etc. This is a process during which, in addition to the choice of procurement method, there are also choices relating to the bidding documentation, the procedures used during the bidding process, the bid-evaluation criteria that will form the basis for the contractor's choice, the division of project risks and options for risk reduction, *price mechanisms* imposed in the contracts, conditions of payment for works, as well as other conditions that affect the price of design or building and should therefore be identified prior to the invitation of bids. It is complicated work for the owner, who will not be a professional builder. For this reason, at this stage of the building's life cycle, impartial advice from professionals is needed. From the results of procurement come not only the lowest price, but also certainty of price guarantee throughout the construction period in addition to confidence in the specified dates, assurance of the required quality as well as the allocation of risks between the client and the contractor, bearing in mind that any reduction of risk usually costs money. The client's lack of expertise in such matters is the reason why, in the last few decades, a new type of management service has emerged in the construction market, offering construction-client services for the owner – *professional construction management.*

Because the chosen procurement method affects the risk to all contracting parties, and therefore the cost of the contract, the **details** must be specified in the *bidding invitation documents.* The choice of the procurement-method details (price-formation method, bidding procedures, the qualification process for bidders, contract conditions affecting price, etc.) should be made directly before the beginning of the design or construction stages.

In the current chapter, methods and procedures of procurement are described in general. Their systematic implementation during the project's development, through development of the procurement conditions, is different in the design stage and in the construction stage, and therefore they are dealt with separately in the following chapters on design and construction (Chapters 3 and 4), regardless of whether these procedures are conducted by the owner himself, in the role of client, or by the company engaged to perform professional construction-management services for the owner.

Goal

The goal is to choose a procurement method, which is determined by the contract type (*traditional contract, design and build contract, professional construction-management contract,* their subtypes and combinations), the *contract's price mechanism* (*fixed price, cost-reimbursement price* or sub-types), method of tendering and negotiation (e.g. *partnership* scheme). The choice depends on the objectives and the priorities of these objectives, set in place in the project-determination phase (see Section 1.1). The extent to which the owner wants to participate in project management during the design and construction stages will also have an influence. After the preparation stage has been completed, these objectives can be revised, based on the results of profitability and feasibility studies, detailed area planning and financial conditions.

Executor

The owner himself can choose the method of procurement (contract type); however, it is recommended that he seeks assistance from his *principal contract adviser.*

2.1.1   Traditional method (General Contract)

According to the traditional method, the client enters into contracts separately with the design company and with the building company. When using the traditional method, the building design should be completed at the level of *basic design* before *competitive bidding* or negotiations with the builder begin. Usually the general contractor takes responsibility for the *fixed, lump sum price* of construction, meaning that the difference between the price and the cost is the general contractor's profit. Alternatively, it is possible to use the *cost-reimbursement price* mechanism.

Reasons to choose a traditional method with General Contract

❑ Contract type has been tested in the long term and is the most widely used procurement method.

❑ Ensures largest degree of price guarantee.

❑ Ensures the client of the lowest price if competitive bidding is used.

❑ Ensures balanced allocation of all risks between the client and the contractor.

❏ Client is not obliged to participate in construction management.

❏ Together with the *bills of quantities* (and *unit prices*) and the *specifications* attached to the building contract, it gives the best guarantee of the required quality.

❏ This contract type makes project development and the appraisal of the effects of change relatively easy, and also minimises the number of contract disputes.

The disadvantages of the traditional procurement method are:

❏ The subsequent design and construction stages have, relatively, the longest total duration.

❏ It opens up the possibility for the *main contractor* to use subcontractors who are 'as cheap as possible'.

❏ Although the appraisal of change is relatively easy, the procedures of change to the project are relatively complicated.

Variants

❏ Traditional method:

- procurement for construction starts when *basic design* is completed together with the *bills of quantities* and *specifications*;

- procurement for construction starts when *basic design* is completed without the *bills of quantities* and *specifications*; the contractors should produce the estimated bills of quantities during the bidding process.

❑ Accelerated traditional method:

- procurement for construction starts when the preliminary design is completed together with the approximate bills of quantities;

- procurement for construction starts when the preliminary design is completed without the approximate bills of quantities;

- procurement for construction starts when the scheme design is completed and the builder is involved in design according to previous partnership relationships.

❑ Traditional method based on *cost reimbursement*, which enables an early start to the project. Used for accident or repair works. Makes it possible to involve the builders in design, to use subcontractors who are highly qualified and to change the project more easily. The traditional general contract based on cost reimbursement ensures the shortest duration of the design and construction processes. The client's need to participate in building management is greater than with other variants, as the client must control all construction-cost documents.

With the traditional method, it should be borne in mind that the lowest cost for the project (building) is ensured through **building cost control during the design stage**, and through details of the tendering procedures.

The best-quality outcome is ensured for the building through the lump sum price, and if the basis of price bidding is basic design together with the *bills of quantities* and *descriptions of works (specifications)*, then the latter are the more significant element of the bidding-invitation documents.

If the probability of change to the project is great, then in order to ensure simplicity when implementing the variations, the fixed, lump sum bidding enquiry could include a requirement for the contractor either to submit the estimated bills of quantities together with corresponding *unit prices*, or to submit only a separate list of unit prices. The unit prices would be used only for variations to the project.

2.1.2   Procurement method based on design and build contract

In a *design and build contract* the client enters into a contract with one contractor for both design and building works. The contractor takes responsibility for the project's cost in the early stages. If the intention is to find a design and build contractor through *competitive bidding*, then the *scheme design* should be included in the bid-invitation documents in order that all the bidders have equal understanding of the requirements. If the intention is to engage a contractor through direct negotiations, then it would be a good idea for the client to complete the needs analyses (see Section 1.3), from which further project development would take place with the involvement of the contractor.

Reasons to choose a design and build contract

❑ The reason could be the peculiarity of the project:

- for example, the design and construction of large, technological projects like power stations, water-purification plants, highway gas pipelines, chemical factories, etc.; the existence of design and build contracting companies specialising in the required field;

- when a project could be undertaken using various building technologies and, because of this, a design more detailed than the scheme design could restrict the use of other possible technologies, and, at the same time, it is possible to define the outcome of the project before the start and to produce measurements after employing the contractor;

- when buildings are constructed for speculative purposes, such as warehouses and living space for sale and therefore the buyer is unknown at the design stage;

- in cases of very simple renovation projects; these projects proceed from the need to change utilisation functions or comfort requirements; they do not include complex technical and technological reconstruction.

❑ It is convenient to communicate with only one contractor.

❑ The price paid by the owner is fixed at an early stage in the project.

❑ The designer and builder have previous experience with the building type in question, and their collaboration with the client brings additional expertise in technical and technological solutions.

❑ Shorter total duration of design and construction processes than in the traditional procurement method.

❑ The owner can delegate the procedure for applying for permission to build and to use the building to the contractor, based on the details of the procurement contract.

❑ The disadvantages of this method are the high costs of any variations to the project during construction, and potential problems with the resulting quality.

As such, this contract is entered into during the very early stages of the project, and the outcome might not meet the expectations of an inexperienced owner. To avoid this situation, the owner should seek the assistance of a professional construction manager, or, alternatively, the principal procurement adviser should be experienced in the relevant building type. As the unsuccessful bidders' design costs during competitive bidding are relatively high, it is logical that these companies include them in their next bid and thus the average price level when using this method is probably higher than in other methods. To ensure quality, design documents should be approved by the owner's team before they are used in construction and this, in turn, might lead to a greater expense in this element for the owner. It is usual to use a two-stage prequalification method with a design and build contract.

Variants

❏ The traditional variant, described above, through competitive bidding or through direct negotiations.

❏ 'Develop and build procurement', wherein the scope of design is established prior to the engagement of a design and build contractor, thus dictating the necessary initial data for further building design.

❏ 'Package procurement', in which the contractor uses technical and technological building systems of which he has experience. The determination of the client's project and the needs analyses are focused in this case on alterations to previous drafts.

2.1.3 Procurement method based on professional construction-management contract

The professional construction-management contract is a mandatory contract used when the owner is short of time or resources, or lacks the ability to perform the function of a building client. It is suggested not to use the term *project management* as meaning professional construction management. While this usage is not wrong, during development the same project-management functions are performed simultaneously in the owner's company, in the design company, and in the building company. The professional construction manager does not design or build using his own labour force. In contrast to the building's general contractor or to a design and build contractor, a professional construction manager does not take responsibility for fixed, lump sum building costs, nor does he take responsibility for the building's completion date. A professional construction manager is paid a service fee of an agreed total amount, or, alternatively, by reimbursement of his actual costs. Typically, the construction-management service fee forms 2–6% of the building's cost, depending on the size and complexity of the project (the bigger the building, the lower the fee as a percentage). The selection of a professional construction manager should be determined by his qualifications and experience rather than the amount of his service fee. The professional construction manager usually engages construction contractors through competitive bidding. For the owner the building price (leaving aside the possibility of extra work) is the sum of the bidding prices of all the various contractors, a total that becomes evident after the last contract has been signed. The owner should request approximate building-cost appraisals from an independent quantity surveyor during the preparation and the design stages. A professional-management contract enables work to start in the early phases of the project's development.

Reasons to choose professional-management contracts

❑ The needs (see Section 1.3) of the owner are not sufficiently formulated, and the probability of change so big, that a general building contractor would not take responsibility for the fixed, lump sum price or for the building's time programme.

❑ Time constraints outweigh financial considerations.

❑ Makes possible minimum total duration due to design and construction being combined.

❑ Possible independent evaluation of cost and duration.

❑ The opportunity to use the experience of a building contractor and professional construction manager in the design phase.

❑ Professional construction manager's continuous coordination of designers', builders' and the owner's own activities through the management team.

❑ The disadvantage is that, at the time the management contract is signed, the owner has no guarantee of the project's price and completion date. To ensure that the price is competitive, the owner should require the management contractor to select building contractors using the tendering process (underbidding), and, if necessary, to implement a method with pre-qualification.

❑ Flexibility during design and building makes it easier to make changes in the project, although with larger costs than in the general contract.

❑ The owner's total risk for cost, quality and time is greater than in the general contract or in the design and build contract.

Variants

Variations to the professional-management contract depend on the scope of the project, the required contract-management scheme, the accounting scheme and the building site's work-management scheme.

❏ The scope of project management is determined by which of the project's development stages are involved in the management contract, as follows:

- preparation stage including needs and profitability analyses, preparation of land purchase, preparation of financial schemes, development of the detailed planning, feasibility study, production of the scheme design, design and construction; in short, the owner can delegate all the building client's activities to the management contractor, apart from the procedures of decision-making and approval;

- procurement stage including choice of procurement scheme in detail, choice of the price mechanism, competitive bidding and the preparation of procurement contracts;

- design stage in all phases: preliminary design, basic design, working drawings, even the procedure for permission to build; for more details, see the client's activities in Chapter 3;

- stage of construction including the planning of the construction site and temporary works, conducting the contractors' work and control of *working drawings* and *performance drawings*, acceptance of works, accounting with contractors, owner's supervision, acquiring the permit to use the building, taking over the building, control of the guarantee-period works, etc.

❑ From the viewpoint of the business type of manager, the professional construction manager could be:

- a professional construction-management company;

- if the cost-reimbursement method is used to determine the building price (possibly together with the *target price*), the manager should be a general contractor's company capable of doing part of the works with its own labour force; in this case, an independent owner's supervisor is needed, and the use of a fixed, lump sum should be excluded, as by matching the general contract with the management contract a conflict of interests arises;

- a design company that also completes the design, starting from the ordinary needs analysis and finishing with the management of building works; this method is a combination of the *design and build contract* and the *management contract*; this scheme can be used when there is no doubt that the design company has the necessary competence for construction project management.

❑ There are two possible options for the accounting scheme:

- the professional construction manager prepares agreements with the building contractors, although the agreements are signed by the owner who also pays the invoices submitted by building contractors and accepted by the professional construction manager;

- the professional construction manager also pays the builders' invoices, passing them in turn to the owner; the latter scheme ensures more control over changes and costs, as all information is communicated through a single person: the professional building manager.

❑ From the viewpoint of the construction site's work management, there are two options:

- classic or so-called 'pure' management contract, in which the professional construction manager implements the work of the client's project manager and manages the contractors on the building site (as the manager of the subcontractors in the case of the traditional procurement scheme), calculates the calendar plan and monitors its achievement. He also designs the construction site plan and temporary works, arranges the establishment of temporary buildings and facilities and takes responsibility for safety on the construction site. In this case the owner's supervision should be delegated to an independent person.

- the professional construction manager delegates the management of contractors on site to a *general contractor* type of builder, from whom building-management work on the site is procured in addition to the actual building work using that company's own labour. Delegated management work includes coordinating the work of all contractors on the site, designing the site plan, establishing temporary buildings and the ensurance of safety. Contracts signed with other building contractors must reflect their dual responsibility to the professional construction manager and to the site manager. This scenario does not produce general contractors and subcontractors in the traditional sense of the words.

The required type of procurement can be chosen by combining the options listed above, which represent only the principal variants. The aim of combining variants is to provide maximum adherence to the owner's cost, duration and quality goals. The number of possible combinations is large, so the contract *principal adviser's* advice should be sought when making the decision.

Activities after the choice of procurement scheme

Depending on whether the project's development is at the design or building-preparation stage, the following activity is the decision as to the method of choosing a contractor: either by competitive bidding or by negotiation.

2.2 Methods for choosing the contractor

Preceding procedure

The procurement method and the price mechanism, which will be implemented through contracts, are chosen.

Goal

The goal is to choose a method with which to select the most suitable contractor for the project, depending on the relative importance of the project's cost, duration and quality goals, and also by considering the project's scope and its complexity, previous experience of contractors and possible *partnership* relationships, the expediency, or perhaps necessity, of competitive bidding. In the most general terms, it is possible to select the contractor using two different methods – competitive bidding or negotiation.

2.2.1   Competitive bidding

Usually the building contractor is chosen through *competitive bidding – tender* (underbidding). With a design and build contract the design work is part of the tender. If the search is for a design contractor only, then the use of underbidding, especially on large and complex projects, should be the subject of careful consideration – is the small economy on the design costs (which are 5–15% of construction costs) more important than the designer's experience and qualifications? Rules and conditions for public procurement are determined by law and their corresponding implementation methods are not the subject of this manual.

Variants

❑ Open single-stage competitive bidding, wherein the bidding invitation is advertised in newspapers or in specialist trade publications. Any interested party can submit a bid.

❑ Open two-stage competitive bidding. In the first stage, contractors are approached to establish whether they wish to participate in the bidding process with the request that they submit documents demonstrating their experience, qualifications and financial capability, or any other necessary qualities. The client then uses this information to pre-qualify the applicants. In the second stage, those applicants who qualify are invited to bid.

❑ Competitive bidding with limited participation, wherein tendering is invited only from previously chosen contractors. The list of contractors recommended by the Association of Construction Clients can be used.

❑ Competitive dialogue is a procurement method in which interested parties can submit applications for participation in the procurement process; the client negotiates with chosen applicants in order that he can identify one or more solutions that suit his requirements. The client makes a request to submit a bid to those potential contractors who participated in negotiations, and chooses a successful bid based on the bidding-evaluation criteria as outlined in the invitation documents. Competitive dialogue is actually a combination of *competitive bidding* and negotiation; in the case of competitive bidding, negotiations are not held before the announcement of the successful bidder in order to ensure the equal treatment of all applicants.

2.2.2 *Negotiations, including partnership*

The bid is invited from a single contractor when:

❏ Only one company is capable of doing the work.

❏ In cases of small building works when the process of bidding can become more expensive than the savings achieved by the result.

❏ If there is an emergency situation and there is insufficient time for the arrangement of competitive bidding.

❏ If competitive bidding does not produce results.

Conducting price negotiations with a single contractor assumes that the client has the necessary ability. In such cases, it is prudent to seek the assistance of the principal procurement adviser. Negotiations enable the involvement of the building contractor in the early stages of the project's development and this in turn allows the building contractor's experience in design and cost formulation to be used (*tailor made price*). Partnership as a procurement method arises from this type of professional relationship. Partnership is based on the mutual trust of contracting parties stemming from the experiences of earlier collaboration, creating a working environment in which all parties benefit (a win–win environment).

In cases where a contract is based on a one-project *partnership*, it is possible to save 7% of the owner's costs, and in cases of a multi-project partnership, the amount saved can reach 15%. It is expected in partnership situations that the designer will be involved in the building contract agreement. In this case the design cost could be higher than average because greater effectiveness is achieved as a result of a multi-option design. Partnership is usually most effective in cases of *design and build* contract management. Partnership can also be used between the owner and the designer separately, later involving the building contractor who was successful in the competitive bidding. Partnership can also involve relationships between the main contractor and subcontractors and/or materials' suppliers. In a partnership situation it is expected that the building client has the necessary ability to deal with the relevant tasks. The need to coordinate in-house documentation between partners may arise.

Decisions about the implementation of the partnership should be made at the highest levels of the companies involved. Partnership starts with the signing of the agreement (partnering charter), in which the goals are formulated: completion of the building on time, adherence to the budget, making a satisfactory profit (savings), avoiding work accidents and defects, and attaining a high quality of finished product.

2.3   Process of tendering

Goal

The goal is to select a contractor and to sign a contract.

Activities

In the activities described below, the procedures, which in public procurement are prescribed by law, are not described in detail. However, private companies could follow this pattern as *good construction practice* when making agreements. Described here are the principal procedures only, with which the owner can ensure the targeted cost, quality and duration.

❑ Dividing procurement into different parts as a response to building-employment priorities and financial considerations (for instance, financial aid from subsidy funds depends on the scope of the project).

❑ Compiling *bidding invitation documents*, including the composition of the bidding forms.

❑ Formulating the bid-evaluation criteria: the lowest price or the most advantageous bid. In the latter case, the other considerations for bid evaluation should be explained. If there are other criteria besides the price then the method of weighting them must be determined. The most rational decision would be to use the lowest price as the criterion for selection, clearly establishing at the same time (through the *bills of quantities*, work descriptions and specifications) the quality requirements for all building materials, products and works that would form part of the bid-invitation documents.

❑ Determining the requirements for bid-time, building-time and warranty-time mortgages.

❑ Determining the time for the submission of the bids, the period of validity of the bids, the time of opening the bids, the inclusion of this information in the bidinvitation documents and determining the time for signing the contract.

❑ Advertising the bidding invitation in the press, and in the case of competitive bidding with limited participation, the issuing of invitations.

❑ Opening the bids and bid record-keeping.

❑ Checking the qualifications, financial status, technical and functional ability of the applicants and bidders when the open-bidding method is used.

❑ In cases where the open-bidding method is used, applicants and bidders who do not meet the requirements should be disqualified.

❑ Evaluating bids: making a choice of contractor or rejecting all bids.

❑ Explaining the conditions of the project when requested.

❑ Evaluating alternative tenders, if they were allowable.

❑ Rejecting bids that are, compared with the control estimate prepared during the design stage, unreasonably low.

❑ Notifying all *bidders* of the results of tendering.

❑ Negotiating with the successful bidder to enter into a contract.

Chapter 3
Design stage

Chapter outline

3.1 Preparation phase

3.2 The choice of designer

3.3 Contracting between client and designer (consultant)

3.4 Management of design

Manual of Construction Project Management for Owners and Clients, First Edition. Jüri Sutt.
© 2011 John Wiley & Sons, Ltd. Published 2011 by John Wiley & Sons, Ltd.

3.1   Preparation phase

Preceding procedures

Feasibility studies, detailed planning, completion of the technical requirements for design and the preliminary measures required to connect to the utility networks, acceptance of the investment decision. The reduction of risk in the current stage is possible if the necessary activities described in the project's preparation stage (see Chapter 1) are performed, ending with the scheme design (see Section 1.8) and its approval. It is emphasised once more that in this context *scheme design* does not have the same meaning that it has among designers, but is used here as the owner's scheme for project development. It is certainly possible to design based on less thorough preparation and to leave the search for alternatives and economies to the designer or to the builder, increasing the owner's own risks.

Goal

The goal is to determine design-procurement conditions that will enable the compilation of the building-design documents, and which originate from the goals established during the preparation stage relating to the building's function, quality, cost and time.

Executor

The owner's project manager has the responsibility to procure and manage the design work and, if necessary, seeks the assistance of the contract's principal adviser or professional management company.

Activities

❑ Making the choice of the design-procurement scheme, originating from the preceding choice of the project's general-procurement scheme (see Chapter 2). The most widely used schemes are:

- traditional design contract in which the owner is in a contractual relationship with only one design company, the general designer, who is, depending on the project's character, an architectural or engineering company that involves subcontractors for specialist design work;

- subordinated design procurement, wherein the owner signs an agreement with the general designer as well as with other designers of specific work;

- the owner is in a contractual relationship with a professional management company which is at the same time a general design contractor. The agreements with design contractors are prepared by the professional building manager, although they are signed by the owner. This kind of procurement is also referred to as divided design procurement;

- design and build contract (see Section 2.3).

❑ Choosing the price mechanism that will be implemented in the design agreements.

❑ Revising the scheme design to ensure that all the initial data for starting the detailed and complex design are described.

❑ Examining the *General Conditions of Design* Contract and Standard of Building Design to obtain an overview as to whether they match the present design contract; if necessary, special conditions are formulated and written into the agreement.

❑ Determining the scope of the designer's work.

❑ Determining the design norms, as well as the instructions and standards that do not derive directly from the law, from general conditions of an agreement and from the standard of building's design; or the designer is authorised to make these choices.

❑ Determining the need and the method of implementation of additional studies, e.g. geodetic and geological investigations. These studies can be integrated with design procurement.

❑ Clarifying whether the details of the technical requirements and preliminary agreements, acquired during the feasibility studies for connecting to the utility networks, are valid, e.g. the validity period and the connection requirements. The relevant consultants should be used to evaluate the necessary capacity of resources.

❑ Deciding whether a *ceiling price* of construction should be inserted in the contract for the designer, with the responsibility to adhere to it.

❑ Dividing the project into construction sequences. Design work sequences are derived from this division.

❑ Formulating the design programme and, if necessary, using a consultant:

- choosing the design-procurement scheme and determining the responsibility limits;

- preparing an initial time schedule for design work; if necessary, the division of construction into sequences is taken into account;

- formulating the initial budget for design costs;

- determining requirements and criteria to ensure design quality: the design company should have a quality-management certificate, or expertise is needed for the whole or partial design;

- determining the order and procedures of revision to the project and its stages;

- determining the critical points of the design;

- determining the approval procedures for subsequent design stages.

❑ Identifying the client's requirements regarding document management, use of design data bank and adjusting the coding system for identification of project elements.

The outcome

Choice of the design-procurement scheme; time schedule of design work; conditions for design contract, which should be communicated to all design-work applicants; specified materials that are not required by law but which designers should incorporate as *good construction practice*.

3.2   The choice of designer

Preceding procedures

Completion of engineering and economic data and contract conditions for design.

Goal

The goal is to find the most suitable designer for the current project, taking into account the required technical capability of the designer as well as the ability to control the cost of both the designed building and the design process itself. The order in which goals are presented corresponds to their level of importance.

Executor

The executor of the design work is chosen from within design companies, where the work is done by specialists with competency in the relevant field. According to common law, the basis for competency is a specialised vocational qualification or higher education corresponding to this speciality, together with three years' work experience. Individual family houses, farm buildings, garages and other smaller buildings can be designed by a person without professional qualification, but who has permission from local government for one-off design. The client's project manager deals with the choice of designer, and, when required, involves the principal adviser.

Activities

❑ The choice of procedures, leading in turn to the choice of a designer, can be as follows:

- negotiations with limited number of companies;

- direct negotiations through partnership;

- competitive bidding;

- competition for a design concept leading to the winner's right to design, or a competition without this right.

❑ Composing and completing the bidding-invitation documents:

- bidding invitation;

- project programme;

- the initial design task with the scheme design, if this is completed (the principal contract adviser is involved in composing the initial task);

- conditions of contract affecting the cost of design and its duration;

- a list of documents not generally established but required by the client;

- if the designer is required to be responsible for the building's *ceiling price*, then corresponding conditions should be set.

❑ Choosing the payment method for the design work (price-formation scheme for design); the most commonly used are:

- fixed, lump sum;

- fixed, lump sum together with additional cost calculations;

- cost reimbursement with or without an upper limit.

❑ Compiling the list of desired applicants.

❑ Issuing the *bidding-invitation documents*.

❑ Following the competitive-bidding procedure and/or negotiation.

❑ Choosing the designer, negotiations to sign the contract.

The outcome

Bidding-invitation documents, protocol for opening the bids, protocol for evaluation of bids.

3.3   Contracting between client and designer (consultant)

Goal

The goal is to protect the owner's interests during project development in order to achieve the objectives in cost, time and quality as specified in the conditions of the design contract, taking into consideration that the owner as a non-professional has the weaker position. While drawing up the contract, the client can rely upon *General Conditions of Contract between Owner and Designer*, which becomes binding to the parties involved only when they are referred to in the contract. During contract preparation a sample contract, which is part of the General Conditions, can provide some guidance. For the designing of buildings, the Standard of Building Design helps the client understand the scope and level of detail required for design. If the client wishes the design standard to be followed, this must be referred to in the contract. The Standard of Building Design defines the structure of design, the division of the design process into stages and the scope of every stage. However, referencing the design standard in the contract does not mean that the designer must produce and hand over all the documents listed in the standard. The client must list in the contract all the documents that will be required, for example, the documents listed in the design standard as well as the *QSC calculations*, calendar plan, financial scheme, etc.

It is worth noting that the owner should have the most important conditions in the contract – affecting the cost of building as well as the cost of design – explained to him during the design-preparation phase and that these should be attached to the *bidding-invitation documents*.

Executor

The design contract is prepared by the client's project manager, who has the help of the contract's principal adviser, if necessary. If the scheme of the contract is produced by a design company, then the principal contract adviser should examine it carefully.

Activities

❑ Describing the building sequences of the development project, if necessary.

❑ Correlating the stages and phases of design with the building sequences.

❑ Choosing the design contract's structure together with the formulation of the approval procedure for each stage:

- framework contract plus separate contracts for the design stages;

- a single contract for the whole design, encompassing all stages;

- separate contracts for each design stage.

❑ Preparing the contract text, including the list of required services and documents, the requirements for the form of design documents and the number of copies that will be handed over, requirements for completing design documents in relation to construction procurement, etc.

❏ Completing the contract appendices, i.e. the documents that will be handed over before the contract is signed: the initial design task and/or the scheme project, the outcome of the investigations, permissions and planning requirements, design conditions, specified materials, etc.

❏ Composing the list of additional (non-design) work required from the consultancy company.

❏ Producing the time schedule of the whole *project* (included as a contract appendix).

❏ Composing the list of documents that will be handed over at the completion of the contract and the dates on which they will be handed over.

❏ In order to provide the owner with control of building cost and quality (the required outcome within the budgeted construction cost) during the design stage, a schedule should be prepared showing in which of the design stages the bills of quantities, work descriptions (specifications) and control estimates (*QSC calculations*) will be produced. The requirement for such work should be included in the design contract, as neither the General Conditions of Contract nor the Standard of Building Design classify such work as an automatic activity during the production of the building design. It would be logical to make the *QSC calculations* during the *basic design* stage, as this will bring about a uniformity of the bidding-invitation documents required for procurement. This in turn ensures uniformity within the building contract as well as a clear definition of the scope of work, of the rules for measurement and of the quality requirements. At the same time this opens the way to choose the building contractor through clear and transparent underbidding, avoiding the need to match the requirements for cost and quality using weighting criteria.

❑ The client has an opportunity to include the responsibility to adhere to the *ceiling price* (building cost), set in the previous stages (profitability studies, scheme project), as part of the designer's contract because the owner may by this time be tied to financial commitments (loans, etc.) that cannot be exceeded. The following illustration might help clarify this responsibility: if, during underbidding for building procurement based on the given design, the minimum bidding price were to be more than 10% above the *ceiling price* (which indicates insufficient work from the designer when monitoring building costs), the designer would then be obliged to amend the design solutions and find technical solutions reducing the building cost at his own expense; if the minimum bidding price were to be less than 10% above the *ceiling price*, the design solutions would be amended at the client's expense. It is possible to assign this responsibility to the designer, if, in the preceding design stage, the owner has completed scheme design to the necessary level of complexity, as described in Section 1.8, ensuring the necessary accuracy of cost calculations preceding the design phase.

❑ Arranging a meeting with a prepared agenda in order to ensure mutual understanding of the contract conditions, and the signing of the contract.

❑ Collating the contract documents into one folder and checking their completeness.

❑ In the case of small-scale works, the contract can be replaced with a purchase order, in which case the designer acknowledges the order.

The outcome

A protocol of negotiations with a decision to divide building and designing into sequences and stages, the notifications of additional work and services from the designers, contracts for designing work.

3.4   Management of design

Goal

The goal is to meet the objectives established in the design contract (cost, time and quality) by specifying progressive solutions to the design stages, and, if necessary, amending them. The application for the necessary changes should be recorded as an order, memo or protocol.

The outcome of any changes on the design should be determined as well as on the building cost and construction dates. The client drives design through design meetings.

Executor

The client's project manager or professional building manager is responsible for the management of design work.

Activities

❑ Informing local government of the beginning of design work.

❑ At each stage of design (*preliminary design*, *basic design*, and *working drawings*) and before starting design work, a meeting should be held between the client's project manager, the building's future occupant, the professional building manager, principal contract adviser and the designer, in which the schedule for regular reviews is specified. In the case of partnership procurement, the participation of the building contractor is effective.

❑ In order to achieve the project's functional and quality objectives, the specialised areas of the design should be revised by corresponding specialists. If the project has complex specialised parts, external consultants should be used during the final inspection.

❑ Examining the alternative solutions and making the decision.

❑ For buildings in which there will be a mass gathering of people, the design documents need the approval of experts through compulsory inspections.

❑ When each stage of the design is finalised, the attainment of all objectives relating to that stage is evaluated; any changes made to the initial data or objectives are also updated.

❑ If, in the design contract, the responsibility for the build-ing's *ceiling price* is given to the designer, the most critical points should be stipulated in the design time schedule:

- before starting design work, the general designer assigns building cost limits to the specialist designers, limits that originate from the structural-analogue method;

- after devising the principal solutions to all specialist parts, a phase of integration of design solutions is neces-sary. At this point about 30% of the *preliminary design* phase is complete and this therefore represents the best milestone for cost-calculation control using the same method. If necessary, the project's solutions are revised at this time until they are in line with the given *ceiling price* of the building. It is still relatively easy to amend the project. This is the part of the design phase in which it is common to use the 'brainstorming' approach to find the best alternatives, and consideration should be given to the use of external specialists during these brainstorming sessions. Of course, brainstorming requires additional time (in the case of a larger project with ample preparation time, up to 1 week) and therefore additional expense. During the *basic design* stage the *bills of quantities* must be produced as well as work descriptions and specifications together with the preparation of the control estimate using the *resource estimation method*. Certainly there is an option for the owner not to request QSC calculations at the basic-design stage and confine himself to obtaining the basic-design cost calculations with the same level of detail as used in the *preliminary design*. However, if this is the case, the bills of quantities and work descriptions (specifications) will not be obtained. This would mean in turn that the client leaves the measurement of construction work quantities, to be based on design drawings, to the construction contractors (*bidders*), who might measure them differently (the problem of measurement regulations) and also have different understandings of the quality of the materials or products specified, or, which is worse, not specified at all in the design (giving rise to later disputes over variations to the project).

❑ Approval, stage by stage, of the design documents by the owner, required before starting each subsequent stage of the design work.

❑ Scheduling the calendar plan.

❑ In giving a written agreement to the results of the design stage, the client binds himself to the design solutions. Any subsequent instruction to amend the design could result in a need to change the completion date and also increase the cost. Variation orders are classed as contract documents.

❑ Managing the changes during design must follow the method of treatment of changes agreed upon in the design contract or at the first design meeting. The definition of change is taken to mean that a participant involved in some part of the design process desires to make changes to a design solution from a previous design stage (one that has previously been agreed with all the participants). The principle of the method of treatment is that the party requesting the change explains all the effects that the change will cause. The client must accept any changes before any work on them begins.

❑ By default the approval of the design stage by the owner denotes the approval of the development project's building costs, thus it is obvious that the build-up of cost calculations and a review of them are essential.

❑ The design documents (detailed structural solutions, technical charts and method statements) for the *working drawings* are produced by the construction contractor and approved by the owner's *supervisor*, and, if the supervisor requires, also by the designer (this being one of the supervisor's functions).

❑ All decisions arising from meetings and the results of inspections are recorded. Inspection results are issued to all participants.

The outcome

The design calendar plan, the design-meeting protocols, variation orders, decisions relating to the implementation of changes (including changes to the contract), completed design documents, the acceptance of the design documents, decisions on the approval of design documents in stages, decisions on the approval of building design documents by the owners of utility networks, local government and other departments (those parties with whom adequate precepts were made within the initial design conditions), building permission (if designated in the contract), the general calendar plan scheduled by the client during design work.

The bills of quantities, detailed work descriptions/specifications and control estimate, if designated in the contract.

The owner is entitled to rely on the designer's obligation to retain the building's design documents and initial data and surveying drawings for a minimum of 7 years, or until they are handed over as part of the archive. The owner himself is obliged to retain the building's design documents for the lifetime of the building.

Chapter 4
Construction preparation stage

Manual of Construction Project Management for Owners and Clients, First Edition. Jüri Sutt.
© 2011 John Wiley & Sons, Ltd. Published 2011 by John Wiley & Sons, Ltd.

4.1 Building permit application

The building permit is a prerequisite both for the erection of a new structure and for the extension of existing structures, their rebuilding or demolition. Building permit data will be recorded in The National Building Register. Following receipt of the conditions determined in the building permit is the presumption of obtaining the permit of use. The local authority must issue the building permit, or reject the application, within 20 days from the date of receiving the application. The building permit has no expiry date, except in cases where the construction work does not commence within 2 years, and for this reason the formal registration of the commencement of construction work is relevant. To formally mark this date, the first instance of digging with a spade sometimes takes place.

The outcome

Building permit together with all the underlying documents that sanction its approval.

4.2 Construction procurement programme preparation (preliminary conditions of contract)

The description of the construction procurement programme as a complexity of activities has a dual meaning for the client. First, from a wider perspective, it entails decision-making and the planning of everything related to construction procurement. Second, in a narrower sense, the construction procurement programme is the wording of specific contract conditions significant to the client, conditions that influence the price of the contract. Therefore they must be present in the written part of the bidding-invitation documentation. These conditions must be issued to all *bidders*, as they have a binding significance from the perspective of contract law.

Preliminary activities

The variety of procurement methods described in Chapter 2 allows procurement of construction to be commenced at different stages of project development.

Goal

The goal is to specify the planned procurement scheme by defining the intention and scope of the project (see Section 1.1), the evaluation criteria of the project, determining the general and specific conditions of the building contract that can influence the bidding price, and preparing the documentation for the tender invitation. This stage also includes the preparation of the client's own direct works and supply contracts with third parties (technological equipment for industrial buildings, furniture, fixtures and fittings for the building). The specifications are required in case any changes arise from project development (design documentation is ready) and because of the probable circumstantial changes in the owner's company and the construction market.

Executor

The client's project manager, professional construction manager or principal contract adviser prepares the procurement programme.

Activities

❑ Selecting the method of construction procurement and the bid-assessment criteria (see Section 2.1).

❑ Selecting the price mechanism.

❏ Selecting the price-indexing conditions, usually applied when the duration of construction will exceed 2 years.

❏ Specifying building commencement and completion dates, and, where required, intermediate deadlines and applicable penalty rates.

❏ Determining the warranties and insurances required from contractors for bidding and for the building and guarantee periods.

❏ Revising general conditions of contract to incorporate the variations required.

❏ Specifying the division of the design documents between procurements (subcontracts).

❏ Specifying client's direct procurements (supply contracts and construction direct contracts).

❏ Appointing those subcontractors whose engagement will be obligatory for the *general contractor*.

❏ Documenting the extent of contracts and supplies.

❏ Determining deadlines and rules that will be applied during the process of competitive bidding or negotiation.

❏ The format for the price bid will be created with the goal of guaranteeing the clear comparability of bids. This format will be included with the bidding-invitation documentation.

❏ Where required, the form for the estimated bills of quantities will also be drawn up and this will be described in the bidding-invitation documentation.

❑ Determining whether the bidders will be allowed to submit, in addition to the requested bid complying with tender documentation, any alternative solutions, and, if allowable, the form in which these alternative solutions may be submitted.

❑ Determining the (confidential) upper limit of building cost, above which all price bids will be rejected.

❑ Determining, where estimated bills of quantities are used, the method by which errors will be taken into account when the lump sum price and the total price of the estimated bills of quantities do not coincide. This method is specified in thebidding-invitation documentation.

❑ Determining what documentation or submissions are required from applicants in order to demonstrate their economic and quality capabilities during the competitive-bidding process.

❑ Determining the sequence for taking over the building and its sections, and the payment conditions.

The outcome

Decisions regarding the method of procurement, price mechanism of contract; arrangement of the client's managerial work; division of the project into sequences and the subordination of contracts, decision to determine boundaries between contracts and procurements; method of payment; determination of the ceiling price.

4.3 Choice of contractor

Goal

The goal is to select the contractor who will be most suitable to guarantee the agreed objectives of cost, quality and time.

It should be borne in mind that when setting the cost goals, the lowest price and price guarantee are different concepts, as are the duration of construction and the guarantee of adherence to the given timescale. Regarding the quality criteria, clear requirements should be stipulated in the quality of materials and specific details thereof, as well as in the quality of workmanship, and these must be recorded in the documentation of the design, in the construction-procurement programme and in the contract documentation.

Executor

The client's project manager, the principal adviser of the building procurement or the professional building manager, or a combination of all of them.

Activities

❑ Researching the market (finding the prices of similar buildings and structures, compiling a list of the approved building companies, etc.).

❑ Deciding to choose either the route of competitive bidding or negotiation – follow an alternative described in Chapter 2.

❑ Selecting the evaluation criteria for the bids. If the *basic design* with the bills of quantities and descriptions of works (specifications) is ready at the time of the invitation to tender, it is logical to select the successful contractor according to tender price. In this case the quality requirements have been specified in detail in the design documentation and all the bidders are expected to comply with them. In the case of more stringent requirements in order to minimise the owner's risk, it is advisable to use two-stage tendering with preliminary qualification, although this will increase the duration of the process.

- If construction price is the criterion for bid evaluation, and if the tender-invitation documents anticipate the dates of commencement and/or completion of the building (or any parts), then the client has the right to expect that the bidder accepts these deadlines unconditionally.

- If the client did not include the bills of quantities and specifications (descriptions of quality of workmanship and materials) in the tender-invitation package, he increases his risk with regard to the scope of the project (quantities), quality and time; the client's principal adviser of building procurement should inform him about this. The lowest-price criterion is the most clear and transparent. The determination of selection criteria for the successful contractor by weighting of different criteria increases the risk of corruption, which is present not only in public procurement, but is also possible in the private sector. The client's project manager, who can be either from within the owner's company, or be the principal procurement adviser, or be a professional building-management company, can devise complex criteria by which a favoured contractor might gain an advantage.

- In the case of a design and build contract, it may be advisable to consider the transparently measurable costs and revenue of the maintenance period as part of the evaluation criteria in addition to the building costs. The length of the accountable maintenance period should be stipulated in the tender-invitation documents, thus making it possible to reduce running costs and revenue in comparison to the capital cost.

❑ Completing the bidding documentation:

- tender invitation (by letter or by a press advertisement) and form of tender;

- design documents: description of structure, drawings, descriptions of works, specifications and the bills of quantities; if the estimated bills of quantities are required from the contractor, there should be a reference in the tender invitation as to the rules of quantity measurement and to the standards of construction-cost classifications and forms that the client expects the bidders to follow;

- building procurement programme highlighting any contract conditions that differ from the general conditions of contracts and any contract conditions influencing the price or the tendering-procedure method, etc.

❑ Preparing the order guaranteeing the selected applicants access to tender-invitation documents, copying and paying for these documents; determining the conditions for the tenderers to inspect the nature of the proposed *building site*.

❑ Determining the procedure for competitive bidding (opening the bids, answering questions, additional explanations, etc.) or for negotiation, ending with the acceptance of the most favourable bid, or the rejection of all bids (see alternatives described in Chapter 2). According to *good construction practice*, the selection of the most favourable bid means the client is obliged to enter into a contract with the successful *bidder* in accordance with the conditions specified in the tender-invitation documentation.

❑ Bids with a significantly lower price than the average market price, and/or the price calculated in the control estimate, and lacking any other sound basis for their validity, should be rejected as bluff.

❑ Issuing the protocols of the bid opening and notification of the selection of the successful bidder to all tendering participants.

The outcome

List of contractors approved for competitive bidding with limited participation; the preliminary qualification conditions for contractors; specification of competitive-bidding procedure; protocol for opening the bids; protocol of tender evaluation and decision regarding the selection of the successful bid.

4.4 Construction contracting

Goal

The goal of a building contract is to guarantee that the objectives determined in project development and described in the design documentation and contracting programme are achieved.

A large number of contract clauses will be considered and prepared when making the construction procurement programme and included in the tender-invitation documentation.

The contract documentation (contract with appendices) is essential for the guarantee of the owner's objectives of cost, quality and time. Using the General Conditions of Building Contracts as a contract appendix significantly decreases the workload involved in preparing the contract and generally helps to guarantee that the division of the rights, obligations and responsibilities relating to each of the parties is dealt with according to good construction practice. It is generally not mandatory to use the General Conditions of Contract, and therefore a specific reference to it must be made in the contract in order to add it to the contract documentation. The following paragraphs describe such activities as formulating the contract specifications that are in need of inclusion in the contract according to the General Conditions, or which can be dealt with differently according to the owner's interests. Into the contract documentation also will be included the records of changes made during construction work and the records of construction management meetings. It is advisable to determine in the contract the treatment of changes (the quantitative criteria from which to determine what is a change that will influence the price and completion dates of the project and what is not).

Similar to the case for the owner's general contract, direct contract documents of specific materials' supply, or procurement relating to specific works, will be prepared and matched to the general contract.

Executor

The building contract will be prepared by the client's project manager or professional construction manager, also involving, where required, the principal adviser of the building procurement.

Activities

❑ Integrating General Conditions of Contract with the contract. Ascertaining whether the project requires adjustment stipulations, appending the subordinate clause, '. . . in case it has not been determined otherwise in a specific contract', when required.

❑ Handing over responsibility for activities at the *building site* to the general contractor/building manager.

❑ Preparing the client's direct procurement and supply contracts. In industrial construction these are usually contracts of delivery and mounting connected with the production technology (equipment, inventory, etc.), while in civil engineering they are connected with the interior design (fixed furniture, inventory, etc.). The reason why the client might want to manage the work himself, using direct contracts or his own labour force, might be his greater experience in this field in comparison with that of the general contractor. Integrating the client's own works' schedule into the general calendar plan.

❑ Making a joint calendar plan with the building contractors and designers. The participation of the designer in making a joint calendar plan is inevitable when design and construction work are conducted in a combined timescale.

❑ Determining the scope (level of cover) of risk insurance. The client should study the insurance agreement thoroughly as the insurance policy might contain no comprehensive list of insurance cover.

❑ Specifying the rate of performance guarantee to be used during the construction and warranty periods. Examining the performance-guarantee contract specifications.

❑ Specifying the conditions of advance payment (prepayment) for the client in accordance with the financial plan.

❑ Determining the conditions for building inspection by the owner at the building site (e.g. place of work, etc.).

❑ Determining conditions for the security and safeguarding of the building site.

❑ Appointing a construction work manager and building site manager. In cases where a general contractor is used, he performs these functions. The general contractor draws up schedules of labour by stages (earthworks, erection of structures, etc.) and by subcontractors, and later draws up detailed calendar plans for every stage.

❑ If appropriate, specifying the circumstances in which the contractor will be allowed to make any changes to the project without prior consultation with the client.

❑ Determining whether the subordination of management of co-contracting companies falls under the authority of the main contractor or the construction manager.

❑ Making a list of client supplies and determining the deadlines for their delivery.

❑ If required, appointing subcontractors or submitting a request for approval of the list of proposed subcontractors.

❑ If required, compiling a list of the unit prices that will be used when calculating cost variations to the works that were included in the tender as approximate quantities.

❑ With regard to necessary temporary works and temporary structures, specifying the method of paying for them.

❑ Reviewing the way in which accounting for the work will be structured: the form of the source document, deadlines, penalties, etc., as well as the order of delivery and reception of work.

❑ A sample contract, presented as an appendix to the General Conditions of Contract, helps the preparation of the contract.

❑ It is advisable to appoint, through the contract, a person to act as *mediator* to resolve any possible conflicts or disputes that might arise during the contract period, and also to establish the appropriate procedures for dealing with these disputes as well as the method of payment for this service.

❑ Completing the contract documentation: design documents with (estimated) bills of quantities and job descriptions (specifications), calendar plan, schedule of payments, list of subcontractors appointed by the client, specified materials to be used by the contractor and required by the client but which are not mandatory, permit documentation etc.

❑ Establishing the contract-negotiation protocol.

The outcome

The protocol for revising the General Contract Conditions with the resulting enactments of the contract in the text; decision to subordinate the contracts; contract-negotiation protocol; signed contract(s).

Chapter 5
Construction stage

Manual of Construction Project Management for Owners and Clients, First Edition. Jüri Sutt.
© 2011 John Wiley & Sons, Ltd. Published 2011 by John Wiley & Sons, Ltd.

5.1 Construction management

Goal

The goal is to erect the agreed structure in accordance with the requirements stipulated in the contract with regard to costs, deadlines and quality. The client manages the construction with management meetings and by performing the functions of the client and his *supervisor*. This chapter describes the activities of the owner as the client during the construction stage, irrespective of whether he is performing the functions himself or authorises – by mandatory contract to perform part of them – a professional construction manager, an executor of owner's supervision or other consultant. Complex management activities over and above the owner's activities as construction client are as follows: quality management as an owner's supervision activity (see Section 5.2), cost management (see Section 5.3) and management of the construction timescale. The latter is not described in a separate section, but is described recurrently through quality, cost and change management (see Section 5.4), bearing in mind the association of the quality plan and the payment schedule with the general calendar plan of construction work, and taking into account the impact of changes on the duration of construction work.

The division of labour, rights, obligations and responsibilities between the owner and construction supervisor, and between the owner and professional construction manager, has been described according to good construction practice respectively in the *General Conditions of Contract Between the Owner and the General Contractor* and the *General Conditions of Contract Between the Owner and the Professional Construction Manager*, as well as in the *General Conditions of Contract Between the Owner and the Supervisor*.

Executor

The client's project manager or the professional building manager and the owner's supervisor. Both companies and self-employed entrepreneurs can perform the functions of professional construction management and owner's supervision.

In some countries there is a requirement for the official registration or licensing of the building manager and the supervisor appointed by the owner. When building smaller structures (private residences, summer cottages, farmhouses, individual facilities, etc.), the owner can perform the supervision himself according to the law, nevertheless, the use of specialist assistance is advised. With larger structures, the supervisors of specific facilities (sewerage, heating, ventilation and cooling, electricity and low-current work, etc.) should be specialists in their fields. The work of specialised supervisors is generally coordinated by the supervisor of the general construction works.

The owner can procure the client's project management and supervision service from a professional construction-management company, or, in the case of a larger project, from several companies. In the case of a smaller company, the project manager and supervisor can be the same person. The designer's supervisor is responsible for ensuring that construction work is performed according to the design drawings and follows the principles of the designer. The design supervisor must record his inspection visits in the building site diary. The *general contractor* grants the design supervisor the opportunity to carry out inspections at times agreed between them.

Activities

❑ Determining the boundaries of tasks shared mutually between the owner's/client's project manager and the owner's supervisor.

❑ Supervisor's exploratory research into design and contract documents.

❑ Inspecting the documents of performance guarantees and risk insurance.

❑ Arranging the meeting to commence the work. Drafting the minutes of this meeting.

❑ Issuing the notification regarding the commencement of construction work to the local authority and work inspection department at least three days before the commencement of work.

❑ Arranging regular site meetings. Adherence to the plan regarding cost, quality and dates will be discussed at the meetings. In cases where there are significant deviations, the decision to eliminate the variations will be adopted. When required, the representative of the designer will be invited to the site meeting. Minutes signed by the authorised representatives of the parties attending these meetings become an integral part of the contract.

The outcome

Record of work-commencement meeting; regular records of construction-management meetings; notification regarding the commencement of work; changes in the calendar plan; financial plan and payment schedule according to the decisions of the management meetings; change orders.

5.2   Owner's supervision and quality management

Goal

The goal is to comply with the quality requirements specified in the contract documentation and governed by standards in order to ensure the stability and fire safety of the structure, safety of life, property, environment and health, safe usage of the structure, acceptable noise level, energy saving and thermal insulation of the structure. This is accomplished using the principle that supervision must prevent the performance of faulty work. The obligation of the owner to ensure adequate supervision is determined by Building Law.

Executor

The executor of the owner's supervision operates as the representative of the client on the *building site*. The owner's supervisor records his observations in the building site diary, signs the act for the covered works and determines the protocols of the required inspections and meetings using minutes and memos. Where the project is smaller, the owner's supervisor can be given additional tasks regarding the cost monitoring of construction and adherence to the time schedule, as well as the core task of ensuring construction according to the legally enacted requirements. Supervision must be carried out from the commencement of construction work until the occupancy permit is obtained. This supervision must be independent from that of the building and design contractors.

Activities

❑ Verifying the design conformity of the building to design requirements, with a right to require that the designer should immediately bring the design documents to conformity, or in the event of any doubt, require that the owner instigates an expert assessment of the design.

❑ Verifying the conformity of the building designer or guarantor to requirements.

❑ Verifying the conformity of materials permanently installed in the structure of the building, as well as the conformity of elements and equipment, and the confirmatory inspection (with a right to require the submission) of the supporting documentation to verify quality: certificates, conformity declarations, guarantee letters, service and maintenance manuals. Arising from the above is the right to require the replacement of any nonconforming items.

❑ Making copies of important documentation verifying the quality; collating the documentation (forming sub-folders) according to the instructions relating to the classification of construction costs.

❑ Inspecting the conformity of the structure under construction with the building design, with the right to require the reconstruction of any nonconforming work.

❑ Inspecting the maintenance of the structure and building site.

❑ Inspecting the safety of the structure and building site, including the site's effect on the safety of the neighbourhood and third parties.

❑ Inspecting partial completion of construction work according to calendar-plan deadlines.

❑ Where there is nonconformity, immediately informing the owner of the situation (in the event of non-compliance with quality requirements, informing within 2 days).

❑ Participating in the examination of construction work required to produce documentation and the approval of intermediate stages and special works.

❑ Participating in the geodetic marking of the structure.

❑ Participating in inspection of covered works and drafting corresponding reports; when required, making the list of covered works that need inspection.

❑ Participating in building site meetings.

❑ Participating in surveying checks.

❑ When required, making the list of working drawings required from the construction contractor. Approving drawings before the commencement of work based on these drawings.

❑ Inspecting maintenance and operating instructions.

❑ Inspecting as-built drawings.

❑ Ensuring that instructions given to contractors, including conditions laid down by public authorities, are met.

❑ Ensuring that the contractor maintains records in the *building site diary* and that he confirms the authenticity of these entries with his signature. Daily entries in the diary should reflect:

- weather conditions;

- completed work, number of operatives on site;

- machinery and equipment used;

- occupational safety (instructions issued, occupational accidents);

- defects;

- changes and delays.

❏ The owner's supervisor presents his notes and suggestions in writing and bearing his signature.

❏ The owner's supervisor ensures compliance with the requirements in force for building and building safety as the general contractor carries out construction.

The work of the owner's supervisor is made easier when he knows which quality-management procedures the contracting company are following. This becomes possible when the quality-management systems used by the general contractor, in professional construction management, and in (general) design companies, are certified (using ISO 9001). The client may require the relevant manual to be submitted by all the contract or service providers during their preliminary qualification, although in reality this provides no guarantee of the quality of the resulting service. Obtaining the certificate has often become a formal goal and, in practice, the procedures for ensuring certified quality may not be followed.

Therefore it would be useful for the owner's supervisor to insist on a construction-quality plan from contractors. The quality plan is a document (matrix) similar to the calendar plan in which structures (elements), specific operations, engineering equipment and communication systems (measures for quality inspection) are displayed in rows and the controlling activities are shown in columns (the implementation of which depends on the readiness of the element or work):

❏ Engineering inspections inviting the owner's supervisor and/or local authority inspector, administrators of engineering networks, representatives of departmental inspectors.

❏ Inspections of covered work.

❑ Tests of structures, materials or equipment.

❑ Building site inspections.

❑ Occupational safety inspections.

❑ Operations relating to periods of heightened risk (work during severe-weather periods, construction of structures with high contamination-resistance or weather-resistance requirements, etc.).

The outcome

Quality plan; building site diary entries; material and product quality certificates and reports of conformity; required protocols of work inspection and product testing; *act of covered works*; change orders; verification of working drawings; as-built drawings.

5.3   Cost control

Goal

The goal is to ensure the accuracy of the costs according to the schedule of payments, ensuring that they are within the limits of the price agreed in the building contract, inspecting quantities and quality of completed works underlying the invoices.

Executor

The owner's supervisor inspects the quantities and the quality of the work included in the invoices for payment, confirming the details with his signature. The client's project manager, responsible for the construction management of all criteria (cost, quality and time), arranges the cost inspection, relating it to the calendar plan and the cost schedule and completing the documentation required for payment. He calculates the impact of changes on the contract price and confirms the payment eligibility of the invoices with his signature. The owner is then responsible for the due payment of invoices.

Activities

❏ Inspecting the quantities of works shown in a particular item of the work's report on the completion invoice, when presented for payment, as well as inspecting its quality.

❏ Filling in the accounting table of approved actual costs. The calculation table must be clearly comparable to the plan of costs (payment schedule) pertaining to the contract, both with regard to the costs of the current period as well as to the cumulative costs for the project as a whole.

❏ Where there is a divergence from the plan, a problem analysis will be carried out and measures to solve the issue will be developed. The owner will be immediately informed if a risk of exceeding the contract price is identified. If required, the building contract, building project, financing plans and loan agreements will be amended.

❏ The conditions for approving the reports of completed works and invoices, along with the payment conditions and payment timescale, or conditions for refusal and penalty rates, are described in detail in the General Conditions of Contract, as are the methods for resolving any disputes that arise.

❏ Filing the reports of completed works into a suitable register.

The labour resources and procedures for the client's and owner's supervision depend significantly on the type of *price mechanism* applied to the construction contract. The following options are possible:

- The contract price has been agreed as a *fixed lump sum* and a list of estimated *bills of quantities* has been added to the contract; in which case a report of work completion will be made at the level of detail corresponding to the *unit price* listings in the bills of quantities, grouped according to the cost groups of the construction cost-classification standard.

- The contract price has been agreed as a *fixed lump sum* based on *approximate bills of quantities*; in which case the report of work completion will be made in a similar way to the point above, but the contract price will be amended according to the quantities specified during the work, using the *unit prices* used in the contract documentation.

- The contract price has been agreed as a *fixed lump sum*, although *no bills of quantities* described at unit price level were added to the contract; in which case, work-completion reports will be made at the same level of detail as the *bill of activities* and be comparable to the list of the payment schedule and calendar plan works.

- Construction price based on *cost reimbursement*; in which case the contractor must present all the cost documentation (invoices for materials, elements, equipment and machinery used, calculations of salaries and wages paid, subcontractor's invoices, etc.). This requires that the client must have competence as a construction quantity surveyor, or that the client procures an invoice-inspection service from a professional construction-estimating company.

- Construction price based on cost reimbursement, but with a *target cost* – a certain limit set in the contract. The costs saved (the economy) against the target cost will be divided between the owner and contractor in a proportion as agreed in the contract, and, in the event that the target cost is exceeded, all reasonable costs will be paid, although the contractor's profit margin, as agreed in the contract, will be reduced. In addition, a gradual target cost can be set with different triggering standards and a maximum price limit for payment by the client.

- In *fixed-price* contracts, the distinction between minor and significant changes must be defined, thus determining the limit above which there is a right to require a variation to the fixed price.

The outcome

Inspected and signed reports of work completion; approved invoices; written explanations regarding payment refusal; price-variation eligibility due to changes; in cost-reimbursement schemes, the reports made by the client's quantity surveyor approving the contractor's cost documentation; specifications of the financial budget and payment schedule.

5.4   Management of changes and additional works

Goal

The goal is to assess the need for additional works and/or changes and take them into account in the building contract. Processing additional work and changes is relatively easy in the event that the contract documentation includes the bills of quantities and job descriptions (quality requirements or speci-fications). It is easier to calculate the impact of changes on the contract price during construction work when the contract documentation includes the list of the *estimated bills of quantities with the list of unit prices*. The absence of the latter makes it difficult for the owner to verify his 'grounded' (it would be more correct to say 'ungrounded') expectations if he relies only on preliminary-design drawings or even on the scheme design.

Executor

Approving additional work or changes is a part of the responsibilities of the client's project manager, who will have been authorised to do so through the building contract (nominated as the representative of the owner). The right to permit minor and urgent changes will also usually be given to the professional construction manager and/or the owner's supervisor. The authority of the professional construction manager and owner's supervisor will be defined in *mandatory contracts* with the owner. Where changes increase the contract price by more than the reserve given in the client's financing plan, or cause the contract deadline to be exceeded, the owner himself has to approve the changes.

Activities

❏ The request for change or additional work must always be made in writing, except for minor urgent verbal instructions from the owner's supervisor, which must later be recorded in the site diary and, if necessary, in the documentation of changes.

❏ Written consent regarding change or additional work must be obtained before commencing the work.

❏ If the client makes the request for change or additional work, then the contractor, accepting the change, will estimate its impact on the contract price and deadline. Only after the client has accepted the price and amended deadline, and the design documentation has been changed to reflect this, can the contractor commence the modification work.

❏ When the request for change is made by the contractor, he will produce an accompanying estimate of changes to the contract price and deadline. The contractor can commence modification work only after the acceptance of the change and corresponding cost.

❏ The amended price will also be reflected in the payment schedule and tables of calculation for actual costs.

❏ All the parties concerned with the change will be informed of the change or additional work.

❏ As payment for changes or additional work, the cost-reimbursement mechanism can be used even if it has not previously been agreed in the contract.

❏ The descriptions of changes or additional work, signed by the authorised representatives of both contract parties, will be considered as appendices to the contract.

❏ The procedures – and time limitations for their realisation – described in the *General Conditions of Contract Between the Owner and the General Contractor* will be followed in the cooperation between the parties involved if the conditions of the current contract do not specify otherwise.

The outcome

The change orders; cost estimates of changes and impact on duration; changes in design documents; minutes of procedure for dealing with disputed changes; documentation accepting the changes (as contract appendices); performance drawings reflecting the changes; amended calendar plan of the work and payment schedule.

5.5 Management of the client's direct contracts

Goal

The goal is to ensure the integration of the *direct contracts* that the client has negotiated with suppliers of special materials and equipment and with special contractors into the general contract.

Executor

In cases where the obligation has not been assigned by the contract to the main contractor (e.g. with a subordination contract), the client's project manager is responsible for the relationship between the client's direct contracts and the general contract of construction (the calendar plan, the financial plan). The owner's supervisor is responsible for approving work (both in terms of the quantity and quality) as well as other works (see Section 5.2). The owner's supervisor inspects the delivery of materials, products, fixed furniture and inventory. Responsibility for the receipt of special equipment and materials, special works and special services lies with the owner's own specialists (technological divisions).

Activities

❑ The client arranges receipt of the materials, products or equipment procured by him, which are to be installed by the contractor according to the building contract, at the building site together with the inspection of their quantities and their certificates of quality or declarations of conformity. He also arranges the temporary storage of these materials at the building site and their delivery to the building contractor. The condition of the delivered materials will be recorded in the deed of conveyance. After delivery, responsibility for the materials will be transferred to the contractor.

❏ With equipment or installation procurement, in cases when the amount of installation work is relatively small and/or the client participates in the installation work himself, the supervision takes place in the usual way.

The outcome

Reports of receipt and any nonconformity of materials and equipment; certificates, declarations of conformity; deeds of conveyance to the contractor; entries in the site diary.

Chapter 6
Take-over stage

Chapter outline

Manual of Construction Project Management for Owners and Clients, First Edition. Jüri Sutt.
© 2011 John Wiley & Sons, Ltd. Published 2011 by John Wiley & Sons, Ltd.

6.1 Revisions of general construction work

Goal

The goal is to guarantee the realisation of work in accordance with the quantities and quality specified in the contract documentation.

Executor

The client's project manager or the professional construction manager is responsible for arranging regular inspections. The building-control department of the local authority will be invited to the inspections of elements of structures as determined by the building permit.

Activities

❑ Constant inspection of work; recording the results in the site diary, or by another written method, and including them in the diary during archiving.

❑ Inspections to ensure compliance with the quality plan (see Section 5.2); the results are recorded in protocols, reports of covered works, etc.

❑ Inspections by the owner's supervisor to ensure the implementation of instructions resulting from previous inspections.

❑ After the inspection of foundations, a measurement drawing will be produced that can be crucial when planning later superstructures.

❑ Checking, prior to the acceptance inspection, the maintenance manuals and service instructions submitted by the contractor.

❑ Inspection may result in a requirement for the client to instigate a technical evaluation.

The outcome

Protocols; reports of covered work; entries in the site diary; expert reports.

6.2 Revisions of engineering systems

Goal

The goal is to guarantee the realisation of work in accordance with the contract documentation, and to check the performance of the technical systems.

Executor

The inspection of engineering systems will be arranged in accordance with the procedure described in Section 5.5. However, when required, a specialist consultant from the relevant technical field will be invited to carry out the inspection.

Activities

❑ Inspecting installed equipment regularly, leading to regular reports on concealed technical systems.

❑ Before the commencement of the building-acceptance process, the inspection plan for the technical systems will be drawn up determining the inspection times and the names of the consultants to be invited.

❑ Checking the contractor's installation drawings.

❑ Checking the maintenance manuals and user instructions issued by the contractor.

❑ User training.

❑ Work experiments.

❑ Regulation of equipment and systems, surveying and issuing of passports (documents verifying conformity with the manufacturer's parameters) by the contractor. Survey-control parameters must be previously determined by the designer.

❑ Tests of complete systems and the coordination of systems, control surveys managed by the client.

❑ Checking the contractor's hand-over documentation.

❑ Acceptance of technical systems after the making good of any defects.

The outcome

Entries in the site diary; equipment passports; technical system inspection reports, user manuals and maintenance instructions; installation drawings; training-implementation documentation; technical system acceptance reports.

6.3　Building take-over

Goal

The goal is the completed construction. The acceptance of the building can take place in two stages: official acceptance and acceptance by the owner.

Executor

> The client's project manager requests that the representatives of all the project's parties participate in the acceptance process.

Activities

> ❑ Deciding to commence the acceptance procedure for the building, based on the documentation relating to the inspection of the building, its parts and technical systems.
>
> ❑ Forming the official reception committee and coordination of its personnel with local government. The usual participants in the reception committee are the local government building supervisor, representatives of the departments involved in the initial conditions for the design and building permit, the general building contractor, the general designer, the owner's supervisor, the construction client (and prospective occupant).
>
> ❑ Official inspecting and drawing up a protocol (report). The local authority will issue an occupancy permit for the building in accordance with the official deed of receipt. Official receipt does not necessarily mean the delivery of construction work by the main contractor and its acceptance by the owner. However, such a condition must be stipulated in the tender-invitation documentation or in the minutes of the building (launch) meeting.
>
> ❑ Taking over by the owner takes place in accordance with procurement, underpinned by previous general construction work and technical systems' inspection results (see Sections 6.1 and 6.2).
>
> ❑ Confirming the conformity of work to the contract. Drawing up a list of defects with their rectification deadlines and conditions. Agreeing follow-up inspections.

❑ Taking over the complete hand-over documentation from the contractor at a reasonable time before the acceptance inspection, in order that the client may study it.

❑ The decision of the owner regarding take-over of the building will be documented in the acceptance-inspection record.

❑ The performance warranty of the contractor during the construction period will be replaced by the performance warranty of the guarantee period.

❑ Transferring the responsibility for the building, as well as its usage and maintenance costs, from the contractor to the owner.

❑ Open-ended continuation of the building's insurance must be guaranteed.

❑ Final settlement of the construction contract will be completed, and following rectification of errors and defects, the last payment will be made.

The outcome

Inspection records; official take-over report on the building; the contract-acceptance report by the owner; the deed of receipt of the contract documentation; record of follow-up inspections; record of final settlement.

6.4   The building's taking for use

Goal

The goal is to deliver a completed building to its occupants and guardians, and prepare them for the use of the building and its engineering systems. Maintenance manuals and user instructions are usually specified by the designer; the maintenance manuals for the structures and installed engineering systems will be submitted to the client by the contractor.

Generally, the building's take-over (see Section 6.3) precedes its taking for use. However, in some circumstances it is possible and useful for the owner to utilise the structure prior to its completion. In such case, the report on the basic completion of the structure, in which the basic completion of construction work is recorded, must be determined as a result of building inspection.

The report of basic completion states that the owner can commence using the structure regardless of the incomplete state of some elements of work, or regardless that some defects are present, and the relevant departments give their consent for this. There is a clear need to formulate in this report the exact boundaries of responsibility and obligation borne by the contracting parties. Also the details of its determination need to be evident.

Executor

The owner is responsible for the preparation for usage, which is arranged by the client's project manager. During the period of training in the usage and maintenance of the building, the contractors and suppliers of equipment will also be in attendance, if required.

Activities

❑ Appointing persons responsible for usage and maintenance; it is useful to do this prior to building acceptance in order that such personnel can be included in the acceptance process.

❏ Drawing-up the usage and maintenance plan covering the entire life cycle of the structure.

❏ Drawing-up the training plan for users and administrators.

❏ Examining instructions, and manuals of use and maintenance, during the training period.

❏ Rectifying defects that prevent user activity, to be undertaken by the contractor or supplier either without delay or during the warranty period.

❏ Transferring responsibility for maintenance from the contractor to the owner.

The outcome

Utilisation and maintenance plan, training plan and documentation, list of defects preventing utilisation.

6.5   Project completion

Goal

The goal is to complete the project, the disbanding of the construction project-management structure created within the owner's company and storage of records of procedures in which experience has been gained; arrange the storage of design and contract documentation.

Executor

The client's project manager is responsible for project-completion operations; the relevant departments of the owner's company ensure the storage of records of procedures and construction documents.

Activities

- ❏ Producing written rules for the owner's company, determining the labour division and responsibility for the storage of documentation and the records of procedures.

- ❏ Making the plan of action for the warranty period: the regulation, inspection and maintenance that will be required, as well as assigning responsibility for these items to the appropriate personnel.

- ❏ Determining the method for the warranty-period inspection.

- ❏ Determining the method for inspections for the longer-term-warranty structures and equipment.

- ❏ If future construction is planned by the client, it is useful to request feedback from the collaborating partners.

- ❏ Holding a project-conclusion meeting.

- ❏ Disbanding the project-management structure.

The outcome

Action plan during the warranty period; the method for storage of design documentation; the method for storage of procurement contract documents (inspection records, reports of covered work, quality certificates of materials and equipment, site diaries, etc.); the protocol of the project's conclusion meeting.

6.6   Warranty period

Goal

The goal is to guarantee, through the warranty-period operations, the project-based functioning of the structure and the rectification of any defects that might emerge.

Executor

The client's construction project manager, or those personnel within the owner's own management structure who have been identified in the warranty-period action plan, or in the project-completion record.

Activities

❑ Checking that errors, defects and any other conditions preventing the utilisation of the structure are rectified.

❑ Inspecting the technical systems (heating, cooling and ventilation systems, etc.) in actual usage conditions.

❑ Supervising the maintenance of structures and technical systems.

❑ Arranging the second stage of training for users and administrators.

❑ Arranging and conducting the final inspection of the warranty period, checking that defects are rectified, arranging a follow-up inspection.

❑ Collating data on maintenance-period costs and on their comparison to projected maintenance costs.

❏ Releasing the guarantee-period warranty.

The outcome

Protocols of inspection during the warranty period and follow-up inspection; list of defects; training documentation.

Appendices

Appendix 1

List of the construction client's principal decisions

1. Decision to invest
2. Preliminary choice of the procurement scheme
3. Approval of the financing scheme
4. Decision to purchase the plot of land
5. Approval of the scheme design
6. Approval of the detailed area plan
7. Approval of the procurement route and price mechanism
8. Signing of the framework contract of the design
9. Signing of the preliminary design contract
10. Approval of the principal decisions of preliminary design
11. Receipt and approval of the documents of preliminary design
12. Signing of the basic design contract
13. Receipt and approval of the basic design documents
14. Beginning the tender or negotiation procedure with the builders
15. Choice of the owner's supervisor
16. Choice of the general contractor
17. Signing of the contract with the general contractor
18. Decisions regarding change to the design
19. Approval of the working drawings
20. Taking-over the building
21. Final settlement with the general contractor
22. Final inspection during warranty period

Manual of Construction Project Management for Owners and Clients, First Edition. Jüri Sutt.
© 2011 John Wiley & Sons, Ltd. Published 2011 by John Wiley & Sons, Ltd.

Appendix 2

List of document folders that should be completed during project development

1. Project-financing contract documents
2. Land-purchase contract documents
3. Real-estate acquisition documents
4. Servitudes' contract documents
5. Conditions and agreements to connect to utilities' networks
6. Alternative solutions for investigations of needs, profitability and feasibility
7. Scheme-design documents
8. Initial conditions for design from local government
9. Register of recommended contractors
10. Geodetic and topographical plans
11. Preliminary assessment of environmental impact
12. Detailed area plan documents
13. The initial task to design, set by the owner
14. Design-contract documents
15. Changes to design contract
16. Preliminary design documents
17. Permit-to-build documents
18. Basic design documents with bills of quantities, specifications and control estimates
19. Minutes of design meetings
20. Invoices for design work
21. Documents of invitation to bid for construction
22. Bidding documents
23. Minutes of tendering meetings
24. Construction-contract documents
25. Changes to construction contract
26. Owner's supervision contract documents
27. Building site diaries
28. Minutes of site meetings
29. Covered works revision acts

30. Quality certificates and declarations of conformity for materials and equipment, grouped according to construction-cost classification standards
31. Construction works take-over documents, grouped in the same way
32. Invoices for construction work
33. Invoices for supervisory work
34. Testing protocols for materials, equipment and technical systems
35. Prescriptions and claims
36. Register of project costs
37. Surveying drawings
38. Documents of revision for technical systems, structures and facilities
39. Documents of building's take-over
40. Documents and minutes from the final settlement meeting
41. Documents permitting use of the building

Bibliography

Ashford J. Management of Quality in Construction. Taylor & Francis e-library, 2003 (original version published by E & FN Spon, an imprint of Chapman & Hall, 1989).

Barrie, D.S., Paulson, B.C. Professional Construction Management, Fifth edition. The McGraw-Hill Companies, Inc., 2006.

Boyd D., Chinvio E. Understanding the Construction Client. Blackwell Publishing Ltd, 2006.

Brandon P., Shu Ling Lu. Clients Driving Innovation. Blackwell Publishing Ltd, 2008.

Briscoe G. The Economics of the Construction Industry. B.T. Batsford Ltd, 1988

Cain, C.T. Profitable Partnering for Lean Construction. Blackwell Publishing Ltd, 2004.

Ferry, D.J., Brandon, P.S. Cost Planning of Buildings, Eighth edition. Blackwell Publishing Ltd, 2007.

Harris, F., McCaffer, R. Modern Construction Management, Sixth edition. Blackwell Publishing Ltd, 2006.

Levy, S.M. Project Management in Construction, Fifth edition. The McGraw-Hill Companies, Inc., 2007.

McGeorge, D., Palmer, A. Construction Management: New Directions, Second edition. Blackwell Science Ltd, a Blackwell Publishing Company, 2002.

Raftery J. Principles of Building Economics, Second edition. Blackwell Science Ltd, a Blackwell Publishing Company, 1998.

Rougvie, A. Project Evaluation and Development. B.T. Batsford, Ltd, 1995.

Turner, A. Building Procurement, Second Edition. Palgrave Macmillan, 1997.

Manual of Construction Project Management for Owners and Clients, First Edition. Jüri Sutt.
© 2011 John Wiley & Sons, Ltd. Published 2011 by John Wiley & Sons, Ltd.

Index

Manual of Construction Project Management for Owners and Clients, First Edition. Jüri Sutt.
© 2011 John Wiley & Sons, Ltd. Published 2011 by John Wiley & Sons, Ltd.

Keep up with critical fields

Would you like to receive up-to-date information on our books, journals and databases in the areas that interest you, direct to your mailbox?

Join the **Wiley e-mail service** - a convenient way to receive updates and exclusive discount offers on products from us.

Simply visit **www.wiley.com/email** and register online

We won't bombard you with emails and we'll only email you with information that's relevant to you. We will ALWAYS respect your e-mail privacy and NEVER sell, rent, or exchange your e-mail address to any outside company. Full details on our privacy policy can be found online.

WILEY-BLACKWELL

www.wiley.com/email

17841